AF539705

CUCURBITS

CUCURBITS

Biotic and Abiotic Stresses

Nripendra Laskar
Bholanath Mondal
Partha Choudhuri

NEW INDIA PUBLISHING AGENCY
New Delhi – 110 034

NEW INDIA PUBLISHING AGENCY
101, Vikas Surya Plaza, CU Block, LSC Market
Pitam Pura, New Delhi 110 034, India
Phone: +91 (11)27 34 17 17 Fax: +91(11) 27 34 16 16
Email: info@nipabooks.com
Web: www.nipabooks.com

Feedback at feedbacks@nipabooks.com

ISBN No. 978-93-86546-50-0

Composed, Designed & Printed in India

Dr. Debabrata Das Gupta
Retired Professor, Palli Siksha Bhavana
Visva-Bharati, Santiniketan, Bolpur
West Bengal-731235, India

Former **Vice Chancellor**
Bidhan Chandra Krishi Viswavidyalaya
Mohanpur, Nadia
West Bengal-741252, India

Foreword

It gives me much pleasure and satisfaction that a book entitled "Cucurbits: Biotic and Abiotic Stresses" authored by Dr. Nripendra Laskar, Dr. Bholanath Mondal and Dr. Partha Choudhuri is being published. Needless to say that, it is a product of sustained effort that has fulfilled the need with commendable distinction.

Cucurbitacae or gourd family is a plant family consisting of about 965 species in around 95 genera, the most important of which are *Cucurbita* (squash, pumpkin, zuchini and some gourds), *Lagenaria* (mostly inedible gourds), *Citrullus* (watermelon), *Cucumis* (cucumber) and *Luffa*. The plants of this family are grown around the tropics and in temperate areas where those with edible fruits are among the earliest cultivated plants both in the Old and New World. The fruits of most of the Cucurbitaceae family are used for human consumption. These are the highly valuable sources of carbohydrates, proteins, minerals and vitamins and thus play pivotal role in balanced nutrition of human beings. Such vegetables are basically worm season crops and fetch several biotic and abiotic stresses that considerably affect yield potentiality of the crops both quantitatively and qualitatively. The former stresses (biotic) include insects, mites, nematodes, parasitic plant and diseases (caused by fungi, bacteria, phytoplasma and viruses) and the later stresses (abiotic) include nutritional imbalance and environmental parameters.

All such biotic as well as abiotic stresses of cucubitaceous vegetables have been narrated with facts and figures in considerable details in this compendium. Undoubtedly, there are multiple reasons to ascertain that such book would be indispensable to teachers, researchers, students of concerned branch of science. Dr. Laskar, Dr. Mondal and Dr. Choudhuri deserve credit for such joint contribution. I am sure this book will be useful to all those who are interested in vegetable crops - the farmers, scientists, extension workers, students and policy makers as well.

I wish all success of this publication.

Debabrata Das Gupta
Santiniketan, Bolpur
West Bengal

Preface

Fortunately, almost all kinds of vegetables are cultivated in India and there remain hardly any vegetables which is not cultivated here. A tremendous number of diverse vegetables, 175 in all are grown in different corners of the country. These are the highly valuable sources of carbohydrates, proteins, minerals and vitamins and thus play an important role in balanced nutrition of human beings. Among them one of the most important group is cucurbits. This is an important and a big group of vegetables.

Basically most of the cucurbits are warm season crops, grown in a particular period of the year. But, due to development of advanced scientific agro-technology and unique agricultural inputs, these are now being grown round the year and thus meeting demand of daily dietary requirement of ever increasing population of our country. Like other vegetables, cucurbits are much more prone to insect pest attack mainly due to tenderness and softness as compared to other crops and virtual absence of resistance characters because of extensive hybridization. The extend of damage varies with plant type, location, damage potential of the pest involved and cropping season. Export of Indian vegetables is very low because of this higher domestic requirement and several other biotic and abiotic limitations in crop production. In many instances even cent percent yield loss reported to have been occurred due to these stresses.

Successful cultivation of vegetables is hampered badly due to several biotic and abiotic stresses. Sustainable management of the pests of vegetables is really a complex issue in view of the tropical and sub-tropical climatic condition, intensive cultivation of high yielding and hybrid cultivars under high fertilisation and irrigated condition. Lack of adequate knowledge with regard to plant nutrition, protection and economic constraints added to the complexity of vegetable production. In the changing climatic condition the pest scenario is changing day by day. The invasive alien species of pests are not uncommon under free trade and frequent travelling of peoples from one country to another. A considerable portion of both pre harvest and post harvest crop losses have been witnessed every year.

In order to overcome these escalating problems, use of conventional agrochemicals are being continuing as one of the most powerful weapons as visualized from the fact that vegetables are grown in less than 3 per cent of the total cultivated area, but consume over 13 per cent of the total pesticides produced in India. Indiscriminate and excessive use of pesticides caused development of resistance, resurgence, replacement of pest and also hazards to the non-target biota and ultimately leads to environmental pollution. This is happening due to inaccurate diagnosis of the problem and lack of knowledge about proper crop management technology.

These necessitated seeking scientific information on management of biotic and abiotic stresses in cucurbit cultivation and re-evaluating the role of agrochemicals. The information on insect pests, mites and nematode parasites, diseases, weeds of cucurbits and their sustainable management are scattered. There is scarcity of books having exhaustive and exclusive information on biotic and abiotic stresses of cucurbits and their eco-friendly management.

This book deals comprehensively on these issues along with their integrated management in easy-to-understand language. Good quality photographs of some of the important insect pests, disease symptoms, weeds etc. have been provided. The book will be useful for the students, researchers, plant protection specialists, extension workers and the growers as well. Suggestions to improve the contents of the book are most welcome.

The publisher, New India Publishing Agency, New Delhi deserves the commendation for their professional contribution.

Authors

Contents

1

Cucurbits

1.1 Introduction

'Cucurbits' is a term coined by Liberty Hyde Bailey for cultivated species of the family Cucurbitacae. During this current century the term has been used not only for cultivated forms, but also for any species of the family Cucurbitacae. Cucurbits are frost sensitive, predominantly tendril-bearing vines which are found in subtropical regions around the globe. A few numbers of species those are native to or cultivated in temperate climate are prolific seed producing annuals or perennials that live for one season until killed by frost (Robinson and Decker-Walters 2004). Other vernaculars applied to the family Cucurbitacae and several of its members are 'gourd', 'melon', 'cucumber', 'squash' and 'pumpkin'.

Among the vegetables, the cucurbits form one of the largest groups with their wide adaptation from arid climates to the humid tropics. The family consists of about 118 genera and 825 species. In Asia, about 23 edible major and minor cucurbits are grown and consumed. They are grown in summer and rainy seasons in India and even in winter in some parts of southern and western India as both annual and perennial crops.

These are also the main constituents in the kitchen garden of the farmers. Of these, only gourds and pumpkins are cooked and consumed as vegetables, fruits of cucumbers are usually taken as salads while musk-melon and water-melon are consumed mainly as desert. These, as well as ash gourd and pointed gourd fruits are also used in making ketchups, sweets and candies (Srivastava and Butani, 2009). Some of them (e.g. bitter gourd) are well known for their unique medicinal properties. Cucurbitaceous vegetables are in general a good source of vitamin A and C and various other vital minerals.

The bottle gourd has been used by many societies in diverse and interesting ways. It probably was first used as a water carrier, but quickly found diverse uses in making pipes, snuff boxes, musical instruments, cricket cages and even life jackets. Containers crafted from the fruit rind of the gourd were in constant use as bottles for carrying wine and water, making the name "bottle gourd" especially appropriate for the crop. A host of musical instruments were also fashioned from bottle gourds, which are in use even today.

1.2 Cucurbit Vegetables Grown In India

1.2.1 Snake gourd, *Trichosanthes anguina* Linn. (2*n*=22)

Centre of origin: Southern and Eastern Asia, Australia and islands of the Western Pacific.

Snake gourd is a commonly grown vegetable in India and widely distributed in humid tropical areas of many countries including South eastern Asia, China, Australia, Japan, Fiji, Mauritius, Java, South America and some parts of Africa (Swarup 2006). It is an annual creeper having very long and slender stems that are furrowed and sub-glabrous. The leaves are 5-8 cm long and of different shapes. Flowers are monoecious. Fruits are half to one meter long, often twisted and appear like snakes hanging on the supports or ground. This subtropical plant grows very fast in warm climates and produces lots of fruits for a long time.

It is suitable for growing in home garden and fresh market. However, it is an important vegetable in Kerala, Tamil Nadu, Andhra Pradesh and Karnataka. It is important as a good sources of minerals, fiber and nutrients to make the food wholesome and healthy (Ahmed *et al.,* 2000). The vegetable is also grown in Bihar, West Bengal, Uttar Pradesh and Maharastra (Swarup, 2006). In south Indian condition the crop is sown in June-July and in northern India during December-January.

1.2.2 Ash gourd, *Benincasa hispida* (Thunb.) Cogn. (2*n*=24)

Centre of origin : South East Asia.

Ash gourd is also known as wax gourd or winter melon. This popular vegetable is cultivated in tropical regions of South Asia, South-East Asia, China, Japan, Central America, Florida and USA. The vegetable is cultivated throughout India including Kerala, Tamil Nadu, Andhra Pradesh, Karnataka, Uttar Pradesh, Rajasthan, Bihar and West Bengal. It is an extensively trailing or climbing herb having five-angled leaves. This viny cucurbit produces big bright yellow flowers. The leaves are long around 10 to 20 cm and have a hairy long stem. The fruits are usually oval in shape and grow up to 30 cm in diameter. The fruits are cultivated mainly for culinary, medicinal or preparation of sweetmeat known as "Petha." The fruit is rich in vitamins and minerals like phosphorus, calcium, riboflavin, iron, thiamine, niacin and Vitamin C. Methanolic extract of ash gourd fruit showed anxiolytic and anti-depressant activity (Rukmani *et al.* 2003). The fruits are broadly cylindrical or spherical in shape and are covered by white, chalky wax, which deters microorganisms and helps to impart an extraordinary longevity to the gourd. Seeds are numerous, much compressed and marginal.

This is warm season cucurbit and cannot tolerate frost. The optimal temperature for the growth of ash gourd is in the range of 24– 31°C. The plants are adapted to a wide range of rainfall conditions. It tolerates a wide range of soil but prefers a well drained, fertile sandy loam soil with a pH between 6-6.5. In southern India ash gourd is sown by seeds during June-July for October harvest and during December-January for April-May harvest. The crop is sown during summer and February-March in northern and eastern Indian condition respectively.

1.2.3 Bitter gourd, *Momordica charantia* Linn. ($2n=22$)

Center of origin: Indo-Burma.

It is one of the most popular vegetables in India.Bitter gourd is an important market vegetable in Southern and Eastern Asia and is widely spread throughout most of tropical Africa (Krawinkel and Keding 2014). Bitter gourd is a monoecious climber with sub-orbicular leaves, 5-7 lobed with yellow and solitary flower. It is cross pollinated in nature and in India pollination is commonly done by bees (Behera, 2004). The fruits are 5-25 cm long, pendulous, fusiform, beaked, ribbed with numerous tubercles and are used in a variety of culinary preparations and posses high nutritive and medicinal value. The fruits are rich in vitamin C and folate, and contain alkaloids likely momordicine, saponine and albuminoides, which are medicinally important. Juice extracted by crushing bitter gourd fruits is most commonly used for treatment of diabetes (Zhu *et al.*, 2014). Leaf and stem decoctions are used in treatment of dysentry, rheumatism, and gout (Subratty *et al.*, 2005). Moreover, the crude protein content (11.420.9 g/kg) of bitter gourd fruits is higher than that of tomato and cucumber (Xiang *et al.*, 2000).

Bitter gourd can be cultivated from low land to altitudes up to 1,000 m. It requires a minimum temperature of 18⁰C during early growth, but optimal temperatures are in the range of 24–27⁰C. The crop can tolerate low temperatures, but extreme cool temperatures will retard growth. The plants are adapted to a wide variety of rainfall conditions. It tolerates a wide range of soil but prefers a well drained sandy loam soil rich in organic matter with a pH between 6-7.5. It grows well on silty soil on river beds. The crop is mostly direct seeded and sown during January/ February-March in Indian condition.

1.2.4 Bottle gourd, *Lagenaria siceraria* (Mol.) Stand. (2*n*=22)

Center of origin: Ethiopia.

Bottle gourd is an important vegetable of Africa and Asia. In India the crop is cultivated in almost all parts of the country both in hills as well as plains. This gourd is now widely cultivated throughout the tropics, especially India, Sri Lanka, Indonesia, Malaysia, the Philippines, China, tropical Africa and South America. Bottle gourd is probably one of humankind's first domesticated vegetable species providing food, medicine and a lot more. Due to its crispy, soft, and tasty fruits, it is equally liked by rich and poor people (Ram *et al.*, 2006). It is a climbing or trailing herb, native to Northern Coast of Peru, where it grew about 5000 years ago. Leaves are simple, cordate or ovate in shape with serrated margin. The stem of bottle gourd are green angular or round in shape having medium or sparse pubescence. The flowers are large, white coloured, solitary and appear on leaf axils. The fruit shows wide variation in shapes, size and colours ranging from long or oblong to round oblong and even club shaped and fruit colour varies from dark green, light green, dark green with white stripes to almost white (Chatterjee and Maitra, 2014).

Bottle gourd requires a hot and moist climate. Bottle gourd grows in a wide range of soils, but prefers well-aerated, fertile soils with pH 6.5-8. The crop can be direct seeded or transplanted grown on raised or flat beds. The seeds are sown during February-March for summer crop and during June-July for rainy season crop.

1.2.5 Cucumber, *Cucumis sativus* L. (2*n*=14)

Center of origin: India.

Cucumber had its origin as well as domestication in India and spread to other parts of the world. Today it has become the fourth important vegetable in the world after tomato, cabbage and onion (Swarup, 2006). Cucumber (*Cucumis sativa* Linn.) is one of the monoecious annual crops in the cucurbitaceae family that has been cultivated by man for over 3,000 years (Adetula and Denton, 2003, Okonmah, 2011). It is mainly used as a salad crop, whereas oriental pickling melon is largely used after cooking. Cucumber is a very good source of vitamins and minerals. The ascorbic acid and caffeic acid contained in cucumber help to reduce skin irritation and swollen (Okonmah, 2011).

Cucumber requires moderate warm temperature and grows well between 18-20^{0}C temperature. The crop is not comfortable under low temperature condition. The plants are adapted to wide range of soil but prefer a well-drained sandy loam to clay soil rich in organic matter and pH between 5.5-6.7. In southern and western regions of India cucumber is grown round the year mainly during June-July, September-October and December-January.

1.2.6 Pumpkin, *Cucurbita moschata* Duchesne ex Poir. (2*n*=40)

Center of origin: Peru and Mexico.

It is one of the favorite vegetables of many people of India and worldwide. Pumpkin is an annual vine or trailing plant and can be cultivated from sea level to high altitudes. It is famous for its edible seeds, fruit and greens (Stovel, 2005). The name pumpkin is originated from "pepon" the Greek word for "large melon." The pumpkin has long running prickly or hairy stems bearing large flaccid leaves, orbicular or reniform in outline with rounded lobes. Fruits are brownish-yellow in colour; when ripe, round or oval in shape and variable in size with soft peduncle. All plant parts are edible, shoots, flowers and fruits are cooked as vegetables. Seeds too are edible. They are commonly consumed as a main course or side dish, and as an important ingredient of pies, soups, stews, and bakery preparations (Doymaz, 2007, Guine *et al.*, 2012). Pumpkin is also popular in traditional medicine for several benefits like anti-diabetic, antihypertensive, antitumor, immune-modulation, antibacterial, anti-hypercholesterolemia, anti-inflammation, anti-allergic (Fu *et al.*, 2006). It is a rich source of potassium and Vitamin A. The bright orange color of pumpkin is an indication of an important antioxidant, beta carotene. Beta-carotene is the precursor of vitamin A in the body, which performs many important functions in overall health.

It requires a optimum temperature in the range of 24–27^{0}C for successful growth and development. The crop can tolerate low temperature. Well drained sandy loam soil with pH between 6-7 is ideal.

1.2.7 Water-melon, *Citrullus lanatus* (2*n*=22)

Center of origin: Tropical Africa.

Water melon is an important monoecious cucurbits generally grown during summer months. The fruit is generally used as dessert with a cooling effect. Red fleshed watermelon varieties are rich source of lycopene (Mandel *et al.,* 2005). Now, much of the commercial supply of watermelons is grown in Russia. In addition to Russia, the leading commercial growers of watermelon include China, Turkey, Iran and the United States. The leaves are rough on both sides; 60–200 mm long and 40–150 mm broad, but usually deeply 3-lobed with the segments again lobed or doubly lobed. The leaf stalks are somewhat hairy and up to 150 mm long. The tendrils are rather robust and usually divided in the upper part. The fruit is usually globose to oblong or ellipsoid, sometimes ovoid, 5–70 cm long and weighing 0.1– 3.0 kg. The seeds are obovate to elliptical, flattened, 0.5–1.5 cm × 0.5–1 cm, smooth, yellow to brown or black, rarely white (Laghetti and Hammer, 2007; Mabberley, 2008).

Watermelon is a warm season crop, which requires dry weather with abundant sunshine for quality fruit production. The crop can be grown on well drained sandy, sandy loam or alluvial soils. A pH range of 6.0-7.0 and temperature range of 24-27°C are considered optimum for the growth of the vines. Cool nights and warm days are ideal for accumulation of sugars in the fruits. The seed germinates best when temperatures are higher than 20°C. High humidity at the time of vegetative growth renders the crop susceptible to various fungal diseases.

1.2.8 Long melon, *Cucumis melo* L. var., *utilissimus*. Dutch. and Full. (2n=24)

Centre of origin: India.

Long melon or kakri, commonly used for salad or cooking is a warm season cucurbit grown over tropical or sub tropical climate. The plant resembles cucumber and cultivated commonly in Bihar, Uttar Pradesh and Punjab during summer months. The tender fruits are eaten raw or cooked for curries. The fruit length varies from 20-100 cm and cylindrical in shape. Color of fruit may be light or dark green covered with soft hairs over skin. Seeds are small and used in confectioneries. The seed kernel is indigenously used in kidney and bladder stones, ulcers in the urinary tract, suppression of urine, jaundice, vitiligo, ascites, chronic fevers, inflammation of the liver and kidney, and in general debility (Baitar 2003).

The crop is sown during February-March for summer crop and October-November for winter crop in river bed. Deep well drained soil with sandy or sandy loam texture is suitable with pH in the range of 5.8 to 7.5.

1.2.9 Musk melon, *Cucumis melo* L. var *reticulates* Ser. (2n=24)

Centre of origin : India.

It is an important cucurbit grown mainly during hot months in north western parts of India. The fruit is usually consumed as desert. The plant is annual climber with large, hairy leaves and spherical,ovoid or elliptic fruit with different attractive colors. Musk melon is low fat cucurbit rich in vitamins and minerals. It has been shown to possess useful medicinal properties such as analgesic, anti-inflammatory, anti-oxidant, anti-ulcer, anti-cancer, anti-microbial, diuretic, anti-diabetic, and anti-infertility activity (Parle, 2011).

Muskmelons are warm-season crop that grows best at average air temperatures between 18 and 24°C. Temperatures above 35°C or below 10°C retard the growth and maturation of the crop. Muskmelon is very sensitive to cold temperatures and even a mild frost can damage the crop. Good sunshine and high heat increases sugar content in muskmelon. The crop grows well and gives good flavour in well drained sandy loam soil with a pH between 6-7.

1.2.10 Ivy gourd or little gourd, *Coccinia grandis* Linn. (2*n*=24)

Centre of origin: India.

Ivy gourd or commonly known as kundru (Wasantwisut and Thara, 2003) is a tropical cucurbit which grows as minor crop over trees, shrubs, fences and other supports. It is also known as baby watermelon, little gourd, gentleman's toes, tindora or gherkin (inaccurately) is a tropical vine. Ivy gourd is a dioecious, perennial with glabrous stems and tuberous roots with angular or lobed leaves. Fruits are smooth, ovoid or elliptical, 25-50 cm in diameter with bright green stripes when immature which turn to bright scarlet in ripe fruits. It is distributed in Africa, tropical Asia, and is commonly found in Pakistan, India and Srilanka. In India it is mainly grown in Karnataka, Tamilnadu, Kerala, Maharashtra, Andhra Pradesh, Gujrat, West Bengal and Telengana. Ivy gourd is rich in β-carotene and also contains a good amount of complex carbohydrates, fibre, and a vast array of vitamins B and minerals (Gautam *et al.*, 2014).

The crop thrives best in warm and humid climate with a favorable temperature range of 20-32^{0}C. Well drained sandy loam soil is best for its cultivation. Yield and quality is affected in heavy soils.

1.2.11 Pointed gourd, *Trichosanthes dioica* Roxb. (2*n*=22)

Centre of origin: India.

Pointed gourd is a tropical vegetable crop with origin in the Indian subcontinent. This perennial climbing cucurbit is generally found in India, Pakistan, Bangladesh, Himalayas to Ceylon. The crop can survive up to an altitude of 1500 m in Assam and Garo hills of Meghalaya (Ram *et al.*, 2002). It is known by the name of '***patal***' in different parts of Bangladesh and Eastern India including West Bengal and is one of the important vegetables. It is a dioecious climber with perennial root stock; leaves are cordate or ovate-oblong and flowers are dioecious. Fruits are globose, oblong, smooth, 5-12 cm. long with light green stripes on young fruits and red on ripe ones. The fruit is the edible part of the plant which is cooked in various ways either alone or in combination with other vegetables or meat. Pointed gourd is rich in vitamin and contains 2.6 mg Na, 83.0 mg K, 1.1 mg Cu, and 17.0 mg S per 100 g edible part. It is reported that pointed gourd possesses the medicinal property of lowering total cholesterol and blood sugar. It also acts as blood purifier, helps in digestion and retard ageing. The paste of pointed gourd roots acts as sedative in high fever (Anjaria *et al.*, 2002).

The crop prefers a well-drained sandy loam soil with good fertility. February-March is the main growing season. Pointed gourd loves to grow under warm and humid climate with a congenial temperature of 30-35^{0}C.

1.2.12 Ridge gourd, *Luffa acutangula* Roxb. (2*n*=26)

Centre of origin: India.

Ridge gourd is another important monoecious tropical cucurbit native to India. It is a large climber with palmately 5-7 angled or lobed leaves. Fruits are 15-30cm. long, cylindrical or club-shaped with ten prominent ribs or ridges. Some small fruited cultivars produce fruits in cluster. The fruit of these species is cultivated and eaten as a vegetable. The fruit must be harvested at a young stage of development to be edible. The vegetable is popular in China and Vietnam. When the fruit is fully ripened it is very fibrous. The fully developed fruit is the source of scrubbing sponge which is used in bathrooms and kitchens as a sponge tool. It has therapeutic properties and is used for extraction of fibres (Swarup, 2006). Ridge gourd is an excellent laxative and blood purifier. The leaves are commonly used for the treatment of cold related eye diseases and leprosy (Bhattacharya, 2000).

Ridge gourd likes sandy loam soil for its growth and fruiting. The crop grows luxuriantly in warm and hot climate with a temperature regime of 25-35 ^{0}C.

1.2.13 Sponge gourd, *Luffa cylindrica* Roem. (2*n*=26)

Centre of origin: India.

Sponge gourd, or vegetable sponge, is a minor cucurbit and mostly grown in sub-tropical regions in countries such as Brazil, China, Korea, Japan and a few from the areas of Central America (Valcineide *et al.*, 2014). In India this crop is grown in neglected areas with little care. The tender fruit is curried as vegetable whereas dried fruits are used as scrubber. Sponge gourd contains appreciable amount of Vitamin A and C. The crop is laxative, blood purifier and cures jaundice, diabetes, etc. Vines of the crop are more or less same as those of ridge gourd except that the flowers have five stamens and the fruits are smooth and cylindrical. The stem is green and pentagonal and grows climbing other physical solid (Lee and Yoo, 2006). Sponge gourd has alternate and palmate leaves which is 13 and 30 cm in length and width respectively and has the acute-end lobe. Flower is yellow in color. Fruit is large cylindrical in shape.

Sponge gourd is warm season crop and likes to grow and flower within 25-30 ^{0}C temperature. Restriction of female flower production and fruit growth is observed at temperature above 35 ^{0}C. Well drained fertile soil rich in humus is ideal for this crop.

1.2.14 Chow-Chow, *Sechium edule* Jacq. (2*n*=26, 28)

Centre of origin: Southern Mexico and Guatemala.

Chow-chow is an underutilized single seeded cucurbit. It is a monoecious, perennial vine crop with tuberous roots. In India, it is widely grown in Madurai and Nilgiri district in Tamil Nadu, Karnataka, West Bengal, Mandi district of Himachal Pradesh and entire North Eastern region mainly Mizoram, Meghalaya, Sikkim, Manipur and Arunachal Pradesh. Mizoram is a leading state both in area and production (Rai *et al.*, 2002). Chow chow fruit can be boiled, baked, or fried as a vegetable and included in sauces, puddings, tarts, and salads. Every part of this plant like fruits, tender shoots, young leaves and the roots are used as vegetable. Low calorific value of fruits makes it suitable for hospital diets/ baby foods and could also supplement to potatoes for diabetic patients (Singh *et al.*, 2015). Chow-chow is packed with several nutrients and vitamins. The fruits and seeds have higher antioxidant activity

(Ordonez *et al.*, 2006) and are rich in several important amino acids. Chow chow is monoecious, viviparous climber and only single-seeded cucurbit. The vines grow up to 10 m long with tenacious tendrils and are trained on the bower system or on the trees. Its leaves are broadly triangulate and ovate -cordate to sub-orbicular in shape along with 10-15 cm long sulcate petioles and 3-5 divided tendrils. The flowers are unisexual, normally pentamerous, coaxillary and with nectaries at the base of the calyx. The fruit is solitary or rarely occurs in pairs, pear shaped, fleshy, and viviparous. The fruits differ in shape, size, colour, surface smoothness and spine density (Rai *et al.*, 2006).

Chow-chow requires moderate temperature (13-21°C), high relative humidity (80-85%), well-distributed annual rainfall (1500-2500 mm) and 12 hours of daylight for successful growth, flowering and fruiting.

1.2.15 Kakrol or Sweet gourd, *Momordica cochinchinensis* Spreng. (2n=28)

Centre of origin: India.

It is a dioecious perennial cucurbit and commonly found in homestead areas of West Bengal, Assam and Southern part of India. It is rich source of ascorbic acid. Kakrol is commonly propagated through roots. Immature fruits are used as vegetable. Fruits are light green to light yellow in color with tough spines. The plant produces big size root tubers with white to light yellow flowers. The fruits of kakrol are used for the treatment of ulcer, piles, sores and obstruction of liver and spleen and seeds are used for chest problems and stimulating urinary discharge (Khulakpam *et al.*, 2015). The crop comes up well in areas with higher temperature, high humidity and rainfall. The crop is grown in virgin land rich in humus.

1.2.16 Mitha Karela, *Cyclanthera pedata* (L.) Schrad (2n=32)

Centre of origin: South America.

C. pedata is a recently introduced cucurbits from West Indies. This is cold tolerant monoecious cucurbit. It is a type of bitter gourd with a beak. The fruits have no bitterness. It is also called slipper gourd. The plant is cultivated type. Immature fruits are used as vegetable. The fruits of the plant contains peptins, galacturonic acid, resins, lipoprotein etc. (Das, 2015). Young shoots and leaves are eaten as greens. *Cyclanthera pedata* has showed anti-inflammatory, hypocholesterolaemic and hypoglycaemic effects. The plant grows up to 5 m long, glabrous. Leaves are alternate, palmately 3–5-foliolate or simple but very deeply lobed.

Flowers are unisexual, regular, 5-merous; male flowers are borne in axillary, 10–20 cm long panicles whereas female flowers are solitary with inferior, 1-celled ovary. Fruit is an indehiscent, obliquely ovoid berry up to 16 cm long, tapering, flattened, white-green, sometimes with soft spines, many-seeded.

In addition to that there are several minor/underutilized cucurbitaceous vegetables, which are grown and consumed by tribals of North-Eastern region of India. These are mainly *Cucumis hystrix, Cucumis trigonus, Luffa graveolens, Momordica macrophylla, Momordica subangulata, Trichosanthes cucumerina, Trichosanthes khasiana, Trichosanthes ovata,* and *Trichosanthes truncasa.*

Like any other crops, these vegetables and fruits are regularly get affected by several biotic and abiotic stresses like pest insects, diseases, non-insect pests (mites, nematodes, parasitic plants), environmental stresses, nutritional disorders etc. These stresses play pivotal role in limiting production as well as quality of these commercial crops.

2

Biotic Stresses of Cucurbits

2.1 Insect Pests

Insects are often a major obstacle to the successful production of cucurbits (York, 1992). They can severely reduce yield, injure or kill plants, spread diseases and adversely affect fruit quality. A number of insect pests, mites and nematodes have been found to infest cucurbitaceous vegetables but fortunately, in India, fruit flies and a few species of beetles are of economic importance; aphids and blister beetles though of regular occurrence, seldom cause severe damage.

Other insect pests reported to infest the cucurbitaceous vegetables are pumpkin caterpillar, *Diaphania indica* (Sounders), snake gourd semilooper, *Plusia peponis* Fab., white fly, *Bemisia tabaci* Genn., stink bugs, *Apspongopus* spp. and flea beetle, *Phyllotreta* spp. etc. These biotic pressures on the crops are increasing day by day due to round-the-year intensive cultivation of photo and thermo insensitive cultivars as well as indiscriminate use of toxic agro-chemicals in the present era of chemicalized agriculture.

Table 1. Pest insects of cucurbitaceous vegetables

S.No.	Common name	Scientific name	Family	Order
1.	Fruit flies	*Bactrocera cucurbitae* (Coq.), *B. ciliatus* (Loew)., *B. zonata* (Sounder)	Tephritidae	Diptera
2.	Pumpkin beetles	*Aulacophora foveicollis* (Lucas.), *A. cincta* (Fab.), *A. lewisii* (Baly.)	Chrysome-lidae	Coleoptera
3.	Hadda beetle	*Henosepilachna vigintioctopunctata* (Fab.)	Coccinellidae	Coleoptera
4.	Pumpkin caterpillar	*Diaphania indica* (Saunders)	Pyralidae	Lepidoptera
5.	Snake gourd semilooper	*Anadevidia (Plusia) peponis* (Fab.)	Noctuidae	Lepidoptera
6.	Bottle gourd plume moth	*Sphenarches caffer* (Zeller)	Pterophoridae	Lepidoptera
7.	Stem gall fly	*Lasioptera falcata* (Felt.)	Cecidom-yiidae	Diptera
8.	Serpentine Leaf miner	*Liriomyza trifolii* (Burgess)	Agromyzidae	Diptera
9.	Stem borer or clear winged moth	*Melittia cucurbitae* West.	Sesiidae	Lepidoptera
10.	Flea beetle	*Phyllotreta cruciferae* (Goeze)	Chrysome-llidae	Coleoptera
11.	Aphids	*Aphis gossypii* Glover	Aphididae	Hemiptera
12.	Stink bugs	*Apspongopus brunneus Thunberg., A. janus* (Fab.), *A. observes* (Fab.)	Pentatomidae	Hemiptera
13.	Leaf-footed plant bug	*Leptoglossus australis* (Fab.)	Coreidae	Hemiptera
14.	Stem boring beetles	*Apomecyna* spp., *A. saltator* (Fab.)	Lamiidae	Coleoptera
15.	Blister beetle	*Mylabris pastulata* Thunberg.	Meloidae	Coleoptera
16.	Melon thrips	*Thrips palmi* Karny	Thripidae	Thysanop-tera
17.	Leafhoppers (Jassids)	*Eutettix phycitis* Dist.	Cicadellidae	Hemiptera
18.	Mirid bugs	*Creontiades palidifer* Walk., *Cyrtopeltis tennis* (Reuter)	Miridae	Hemiptera

19.	Small dark weevil	*Acythopius citrulli* (Marshal)	Curculio-nidae	Coleoptera
20.	Water-melon weevil	*Acythopius citrulli* (Marshal)	Curculionidae	Coleoptera
21.	Gray weevil	*Mylloceros blandus* (Faust)	Curculionidae	Coleoptera

2.1.1 Fruit flies (Diptera: Tephritidae)

Fruit flies (Tephritidae: Diptera) are recognized as a major group of insects of economic and quarantine importance due to the preference of numerous species infesting different fruits and vegetable crops including cucurbits. Several species of fruit flies have been detected to infest the cucurbits in different corners of the world. They are, *Bactrocera cucurbitae* (Coq.), *Bactrocera ciliatus* (Loew), *Bactrocera diversa* (Coq.) *Bactrocera latifrons* (Hendel), *Bactrocera parvulus, Bactrocera zonata* (Saunders), *Bactrocera tau* (Walker), and *Myiopardalis pardalina* (Bigot) (Kapoor, 1970). Among these, the most dominant pest of cucurbitaceous vegetables is melon fly, *Bactrocera cucurbitae* (Coq.).

a) Melon fly, *Bactrocera cucurbitae* (Coq.)

Hosts of commercial importance:

This is a very serious pest of cucurbit crops. It has been recorded from over 125 plants, including members of families other than Cucurbitacae. However, many of those records may have been based on causal observation of adults resting on plants or caught in traps set in non-host trees (White & Elson – Harris, 1992). Based on the extensive surveys carried out in Asia and Hawaii, plants belonging to the family Cucurbitaceae are preferred most (Allwood et al., 1999). Doharey (1983) reported that it infests over 70 host plants, amongst which, fruits of bitter gourd (*Momordica charantia*), muskmelon (*Cucumis melo*), snap melon (*Cucumis melo* var. *momordica*) and snake gourd (*Trichosanthes anguina* and *T. cucumeria*) are the most preferred hosts.

Syed (1970) recorded *B. cucurbitae* from angled luffa (*Luffa acutangula*), balsam-apple (*Momordica balsamina*), bitter gourd (*M. charantia*), colocynth (*Citrullus colocynthis*), cucumber (*Cucumis sativus*- fruit and stem), luffa (*Luffa aegyptiaca*), melon (*C. melo*), pumpkin (*Cucurbita maxima, C. pepo*-fruit and stem), watermelon (*Citrullus lanatus*), wax gourd (*Benincasa hispida*) and white-flowered gourd (*Lagenaria siceraria*). Tan and Lee (1982) listed the following hosts from Malaysia: bitter gourd, cucumber, watermelon and in Thailand it has also been reared from *Momordica cochinchinensis* (Clausen *et al.*, 1965). Drew (1989) recorded it from cucumber, marrow (a variety of *Cucurbita pepo*) and melon in the South Pacific region.

In the Hawaiian Islands, melon fly has been observed feeding on the flowers of the sunflower, Chinease bananas and the juice exuding from sweet corn. Under induced oviposition, McBride and Tanda (1949) reported that broccoli (*Brassica oleracea* var. *capitata*), dry onion (*Allium cepa*), blue field banana (*Musa paradisiaca* sp. *sapientum*), tangerine (*Citrus reticulata*) and longan (*Euphoria longan*) are doubtful hosts of *B. cucurbitae.* The melon fly has a mutually beneficial association with the Orchid, *Bulbophyllum patens*, which produces zingerone. The males pollinate the flowers and acquire the floral essence and store it in the pheromone glands to attract con-specific females.

The hosts of melon fly that require further confirmation include, African horned cucumber (*Cucumis metuliferus*) and pointed gourd (*Trichosanthes dioica*) (Kapoor and Agarwal, 1983). *B. cucurbitae* has been known to develop in stem galls induced by another insect, namely those of *Lasioptera toombi* (Grover) (Diptera: Cecidomyiidae) in ivy gourd (*Coccinia grandis*) (Bhatia and Mahto, 1968). There are some unusual hosts of *B. cucurbitae*, which it probably only attacks under unusual circumstances (White & Elson – Harris, 1992). In Malaysia it can attack the stems of tomato grown in hydroponics and it has been reported as causing severe damage to Kai choy (a variety of *Brassica juncea*) in Hawaii (Hardy, 1948).

Wild hosts

The melon fly has been recorded from several wild species of cucurbits. The major host in Hawaii is balsam apple. Wild host of the pest in Japan is *Diplocyclos palmatus* (Okinawa Prefecture, 1987) and in Pakistan wild plants of balsam apple, cococynth and *Cucumis trigonus* Roxb. (Syed, 1970).

Distribution

The melon fly is distributed all over the world, but India is considered as its native home. It was discovered in Solomon Islands in 1984, and is now widespread in all the provinces, except Makira, Rennell-Bellona and Temotu (Eta, 1985). In the Commonwealth of Northern Mariana Islands, it was detected in 1943 and eradicated by sterile-insect release in 1963. It was detected in Nauru in 1982 and eradicated in 1999 by male annihilation and protein bait spraying, but was re-introduced in 2001 (Dhillon, 2005b).

It has been recorded from Africa, America and Asian region. In Asia where the pest cause serious damage to several commercial crops are Bangladesh, Cambodia, Hong Kong, India, Indonesia, Japan, Malaysia, Myanmar, Nepal, Pakistan, Sri Lanka, Taiwan, Thailand, Vietnam, Phillipines (White & Elson - Harris, 1992).

Identification

Adults: A predominantly orange-brown species; scutum with lateral and medial yellow stripes; with facial spots. Wing pattern is characteristic with complete and dark costal band that expanded apically and almost covering apical part of cells r_{2+3} and r_{4+5}; Cross veins R-M and DM-Cu thickly infuscated; 3 frontal setae; scutellum yellow, male with pectin.

Third instar larvae: Maggots are large, length 9.0-11.00mm; width 1.0-2.0mm., creamy white in colour.

Attractant: Males are attracted in cuelure.

b) Cucurbit fly, *Bactrocera ciliatus* Loew

It is also known as Ethiopian fruit fly or Lesser pumpkin fly.

Hosts of commercial importance:

It is a pest of cucurbit crops (Hancock, 1989). In Nigeria it was recorded from cantaloupe (*Cucumis melo*), cucumber (*C. sativus*), squash (*C. maxima*) and watermelon (*Citrullus lanatus*) by Matanmi (1975). Cucumber was only found to infest by the pest when over-ripe. It has been also reported from cucumber, melon (*Cucumis melo*), pumpkin (*Cucurbita maxima*), snake gourd (*Trichosanthes cucumarina*), squash (*C. pepo*), watermelon and white flowered gourd (*Lagenaria ciceraria*) in Mauritius by Orian and Moutia (1960).

Some non-cucurbit hosts, namely, beans (*Phaseolus* spp.), cotton (*Gossypium* sp.), okra (*Abelmoschus esculentus*), and tomato (*Lycopersicon esculentum*) were listed by Munro (1984), but those were unusual hosts as reported by Hancock (1981). However, Matanmi (1975) examined tomato in an area where cucurbits were heavily infested and found no attack on tomato. Orian and Moutia (1960) also noted very rare attack by the pest on tomato.

Wild hosts

Many species of cucurbitacae e.g. *Momordicha* spp. are the hosts of the pest as reported by Matanmi (1975) and Munro (1984). In Egypt, wild growing colocynth is believed to be the main reservoir host from which crops become infested (Shaheen *et al.*, 1973).

Distribution

It has been recorded from Africa, Atlantic islands, Indian ocean, Middle East and Asia. In Asia the species found to infest cucurbits in Bangladesh, India, Pakistan, Sri Lanka and China.

Identification

Adults: A predominantly orange species with facial spots, 2 scutellar setae, a yellow spot covering most of the katatergite, anatergite orange, mid femur yellow or orange-yellow and wing with a costal band that is expanded apically to form an apical spot. Male are with a pecten and scutum without yellow stripes (White & Elson - Harris, 1992).

Third instar larvae: Maggots are medium to large sized, length 9.0-10.5mm, width 1.5-2.00mm.

Attractant

Not attracted to cuelure, methyl eugenol or vert lure.

c) *Bactrocera diversa* (Coq.)

Hosts of commercial importance:

One of the potential pests of Cucurbitacae which was recorded from the flowers of angled luffa (*Luffa acutangula*), luffa (*Luffa aegyptiaca*), pumpkin (*Cucurbita maxima* and *C. pepo*), and white flowered gourd (*Lagenaria ciceraria*) by Syed (1970). In a cage study by Syed (1970) it was found that *B. diversa* would readily attack, and develop in, cucurbit fruits. Batra (1964) found that cut flowers of white flowered gourd were never attacked, but ovaries and young fruits were. Kapoor and Agarwal (1983) also listed some additional hosts. They are, banana (*Musa paradisiaca*), guava (*Psidium guajava*), java plum (*Syzigium cumini*), mango (*Mangifera indica*), nutmeg (*Myristica sp.*) and sour orange (*Citrus aurantium*). White & Elson - Harris (1992) suspected that none of those records were confirmed by rearing.

Wild hosts

Kapoor and Agarwal (1983) recorded it from *Tabernamontana dichotoma* Roxb. and *Solanum verbascifolium*, but White & Elson - Harris (1992) regarded them as doubtful hosts.

Distribution

Distributed all over the Oriental Asia including China, Southern India, Sri Lanka and Thailand.

Identification

Adults: Scutum predominantly black with lateral yellow stripes (vittae), face of male entirely yellow without facial spots; face of female with a black transverse line above the mouth opening. Males are without pecten.

Attractant

Males are attracted to methyl eugenol.

d) Solanum fruit fly, *Bactrocera latifrons* (Hendel)

Hosts of commercial importance:

It is basically a pest of solanaceous crops (Hardy, 1973). In South Asia it has also been reared from *Solanum nigrum* and *S. torvum* (Drew, R.A.I., 1989). In India and Pakistan it has been reared from egg plant (Clausen *et al.*, 1965). There are some doubtful records from non-solanaceous plants, namely banana (*Musa paradisiaca*), carambola (*Averrhoa carambola*), coffee (*Coffea* sp.), common guava (*Psidium guajava*), mango (*Mangifera indica*), snake gourd (*Trichosanthes cucumerina*) and sweet orange (*C. sinensis*) (Kapoor and Agarwal, 1983 and Satoh *et al.*, 1985).

Wild hosts

The pest has also been recorded from *Solanum indicum, S. sarmentosum, S. verbescifolium* and *S. virginianum* (Hardy, 1973; Syed, 1970; Vargas and Nishida, 1992).

Distribution

Adventive population was discovered in Hawaiian Islands in 1983. In Oriental Asia population of this species was noted from China, India, Laos, Malaysia, Pakistan, Sri Lanka, Taiwan and Thailand (White & Elson - Harris, 1992).

Identification

Adults: Scutum predominantly black with lateral yellow stripes (vittae), facial spots, anterior supra-alar setae, prescutellar acrostichal setae and 2 scutellar setae. Abdomen is orange coloured, costal band of wings extended up to apical spot, male with pecten and female with a distinctive aculeus tip.

Third instar larvae: Maggots are medium sized, length 7.0-8.5mm; width 1.2-1.5mm.

Attractant: Not attracted to cuelure or methyl eugenol.

e. *Bactrocera parvula* (Hendel)

The fly is recorded as "on flowers" of white flowered gourd (*Lagenaria siceraria*) in Taiwan and in India by Philip (1950). However, White & Elson - Harris (1992) opined that it was probably derived from a misidentification of another species. Present in the Asia-Pacific region (Cogan and Munro, 1980).

f. Peach fruit fly, *Bactrocera zonata* (Saunders)

Hosts of commercial importance:

It is a pest of peach (*Prunus persica*) and sugar apple (*Annona squamosa*) in India (Grewal and Malhi, 1987) and common guava (*Psidium guajava*) and mango (*Mangifera indica*) in Pakistan. In Pakistan it has also been reared from apple (*Malus domestica*), bitter gourd (*Momordicha charantia*), date palm (*Phoenix dactylifera*), okra (*Abelmoschus esculentus*), papaya (*Carica papaya*), paradise apple (*Malus pumila*), pomegranate ("*Punica*" *granutum*) etc. (Syed *et al.*, 1970). The authors also examined egg plant, Luffa and tomato fruits but failed to find any *Bactrocera zonata* and they suggested that records from those plants may have been based on mis-identification of *Dacus ciliatus*.

Wild hosts

The pest has also been recorded from the plants of family Euphorbiacae, Lecythidaceae (Kapoor and Agarwal, 1983).

Distribution

The pest has been trapped in North America but it has now been eradicated (Spaugy, 1988). Well distributed in tropical Asia including India, Indonesia (Sumatra), Laos, Sri Lanka, Thailand and Vietnam (White & Elson - Harris, 1992).

Identification

Adults: The fly is a predominantly pale orange brown to red brown species; scutum with lateral yellow stripes; with facial spots. Wing pattern reduced and male with pecten.

Third instar larvae: Maggots are large, length is about 10.0-11.0 mm.

Attractant: Males are attracted to methyl eugenol.

g. *Bactrocera tau* (Walker)

Hosts of commercial importance:

The fly appears to show preference for infesting the fruits of Cucurbitacae. However, it has also been reared from the fruits of several other plant families and it undoubtedly a potential pest species. It is recorded from cucumber (*Cucumis sativus*) and angled luffa (*Luffa acutangula*) from Malaysia (Rohani, 1987). Other south-East Asian hosts are bitter gourd (*Momordicha charantia*), malay apple (*Syzigium malaccense*) and pumpkin (*Cucurbita maxima*) (Drew, 1989). Hardy (1973) listed several other hosts, did not indicate the origin of those records which should therefore be regarded as unconfirmed. The hosts that he recorded are, common guava (*Psidium guajava*), Luffa, *Trichosanthes* spp. etc.

Distribution

The pest is widely distributed in Oriental Asia including Bhutan, Cambodia, China, India, Indonesia (Java, Sulawesi, Sumatra), Laos, Malaysia, Phillipines, Sri Lanka, Thailand and Vietnam (White & Elson – Harris, 1992).

Identification

Adults: Scutum orange brown and marked with black and with lateral and medial yellow stripes (vittae); with facial spots. Wings with a costal band expanded into an apical spot. Males are with pecten.

Third instar larvae: Maggots are of medium sized, length 7.5-9.0mm., width 1.0-1.5mm.

h. *Myiopardalis pardalina* (Bigot)

The fly is sometimes known as 'Baluchistan fly'. It occurs all over Afghanistan, but *M. Pardalina* has spread across the north since 2002, mostly infesting sweet melons and other cucurbits as well.

Hosts of commercial importance

It is an important pest of musk-melon, water-melon and cucumbers in India also particularly in Punjab and Bihar (Nair, 1995).

Biological activities of *M. pardalina*

Eggs are laid by the sexually mature female in tender fruits only. Up to 130 eggs are laid in a single fruit. Incubation period is about 4 days. The maggots feed on the seeds and pulps surrounding them. They may get drowned if they go deep into the juicy pulp. Larval period lasts for about a fortnight. Pupation takes place in soil or under the fruit. Pupal period is 13-14 days in summer and as long as six months in winter months (Nair, 1995). The pupa can survive immersion in water for two to three weeks.

Adult life lasts for three to four weeks when fed on fruit juice. Overwintering takes place in pupal stage from September to April. Adults emerging in April subsist on the honey dew of aphids.

The most active period of the pest is in July when life cycle is completed in the shortest period of one month. Two to three generations are passed before overwintering starts.

Among the aforesaid fruit flies except *M. pardalina* that come under the sub-family Dacinae, *B. cucurbitae* is the most important and comparatively more common and destructive. Eggs, maggots and pupae of different species are looks alike and it is hardly possible to distinguish them. However, the adult flies of different species have some specific distinguishing features which have already been elaborated earlier. Most of them are highly polyphagous having a wide range of host plants. Biology, mode of feeding and nature of damage are more or less similar in all the species. Hence, they are discussed hereunder collectively.

Biology

The female adult flies puncture the soft and tender fruits with their stout and hard ovipositor and lay eggs just beneath the epidermis. They lay 4-10 eggs per puncture each time. Eggs are 1.0 to 1.5 mm long, whitish in colour, elongate-cylindrical in shape, beautifully sculptured, slightly curved and tapering at both ends. A single female can lay about 200 eggs in her life span of 8 to 10 weeks. A puncture made by one female is often used by others also for ovipositing and a single fruit may have more than one puncture made by one or more females (Srivastava and Butani, 2009). Fecundity was recorded as 158.33±80.28 by Choudhary and Patel (2007). The pre-oviposition period of flies fed on cucumbers ranged between 11 to 12 days (Back and Pemberton, 1917). Pre-oviposition and oviposition periods ranged between 10 to 16.3 and 5 to 15 days, respectively, and the females live longer (21.7 to 32.7 days) than the males (15.0 to 28.5 days) as revealed by Koul and Bhagat (1994). The egg incubation period of melon fly has been reported to be 4.0 to 4.2 days on pumpkin (Doharey, 1983), 1.1 to 1.8 days on bitter gourd, cucumber and sponge gourd (Gupta and Verma, 1995) and 1.0 to 5.1 days on bitter gourd (Koul and Bhagat, 1994; Hollingsworth *et al.*, 1997). On hatching the maggots feed from within the fruit and they undergo three larval instars before going to pupate. The larval period lasts for 3.0 to 21 days (Renjhan, 1949) depending on temperature and host. On different cucurbit species, the larval period varies from 3.0 to 6.0 days as reported by Chawla (1966) and Chellaiah (1970). After completion of the larval period the mature third instar maggots come out from the fallen fruits and pupate in soil. The larvae pupate in the soil at a depth of 0.5 to 15 cm. The depth up to which the larvae move in the soil for pupation, and survival depend on soil texture and moisture (Jackson *et al.*, 1998; Pandey and Misra, 1999).

When fruits do not fall, the maggots either pupate inside the fruit (although it is not usual) or come out and drop down in soil for pupation. Pupae are 5-8mm. long, barrel-shaped and brown to ochraceous in colour. Type of pupa in this case is known as puparium. Pupal period is about 6-9 days in summer and for about four weeks during the months of December January (Nair, 1995). Doharey (1983) observed that the pupal period lasts for 7 days on bitter gourd and 7.2 days on pumpkin and squash gourd at 27 ± 1° C. In general, the pupal period lasts for 6 to 9 days during the rainy season, and 15 days during the winter (Narayanan and Batra, 1960). Depending on temperature and the host, the pupal period may vary from 7 to 13 days (Hollingsworth *et al.*, 1997). On different hosts, the pupal period varies from 7.7 to 9.4 days on bitter gourd, cucumber, and sponge gourd (Gupta and Verma, 1995), and 6.5 to 21.8 days on bottle gourd (Koul and Bhagat, 1994; Khan *et al.*, 1993). Adult life lasts for up to 56 days in male and 66 days n female when fed. When starved the fly does not survive for more than two days. The adults survive for 27.5, 30.71 and 30.66 days at 27 ± 1° C on pumpkin, squash gourd and bitter gourd respectively (Doharey, 1983). Khan *et al.* (1993) reported that the males and females survived for 65 to 249 days and 27.5 to 133.5 days respectively. Overwintering takes place in pupal stage during cool months of the year from September to April.

Mode of infestation and symptoms of damage

Damage is caused by the adult fly puncturing the fruits and by the maggots feeding and tunnelling within the fruit pulp, but at times, also feed on flowers, and stems. Generally, the females prefer to lay the eggs in soft tender fruit tissues by piercing them with the ovipositor. A watery fluid oozes from the puncture, which becomes slightly concave with seepage of fluid, and transforms into a brown resinous deposit. Sometimes pseudo-punctures (punctures without eggs) have also been observed on the fruit skin. These punctures also serve as an entrance for various microorganisms and as a result of this and after egg hatching, the maggots bore into the pulp tissue and make the feeding galleries. The fruit subsequently rots or becomes distorted. Young larvae leave the necrotic region and move to healthy tissue, where they often introduce various pathogens and hasten fruit decomposition and fall off prematurely from the plant. These fruits are not fit for human consumption and reduce the market value. The vinegar fly, *Drosophilla melanogaster* has also been observed to lay eggs on the fruits infested by melon fly, and acts as a scavenger (Dhillon *et al.*, 2005b). There are three larval instars completed within the fruits. The full-grown larvae come out of the fruit by making one or two exit holes for pupation in the soil.

Up to 20% loss in yield may occur during peak infestation period in July (Nair, 1995). However, Dhillon (2005a) stated the extent of losses from 30 to 100%, depending on the cucurbit species and the season. Fruit infestation by melon fruit fly in bitter gourd has been reported to vary from 41 to 89% (Lall and Sinha, 1959; Narayanan and Batra, 1960; Kushwaha *et al.*, 1973; Gupta and Verma, 1978; Rabindranath and Pillai, 1986). Singh *et al.* (2000) reported 31.27% damage on bitter gourd and 28.55% on watermelon in India. The melon fruit fly has been reported to infest 95% of bitter gourd fruits in Papua (New Guinea), and 90% snake gourd and 60 to 87% pumpkin fruits in Solomon Islands (Hollingsworth *et al.*, 1997).

Seasonal abundance

The melon fruit fly remains active throughout the year on one or the other host. During the severe winter months, they hide and huddle together under dried leaves of bushes and trees. During the hot and dry season, the flies take shelter under humid and shady places and feed on honeydew of aphids infesting the fruit trees.

In India, the activity of melon fruit fly, *B. cucurbitae* was high in April-May, when the weather remained warm (26.2-38.0 °C) and humid (90% relative humidity) as stated by Patnaik *et al.* (2004). Population fluctuation of the fly in relation to abiotic factors (relative humidity, rainfall and temperature) was studied by Banerji *et al.* (2005) and found that the infestation was highest during *kharif* season followed by summer and *rabi* seasons. The percentage fruit infestation was positively correlated with minimum temperature during *rabi* and summer seasons. Highest yield was recorded in the summer sown crop (March) with 27.6% infestation followed by *kharif* sown crop (June). Though *rabi* season crop (October) recorded minimum fruit fly infestation but the yield was significantly low. Population fluctuation of *Bactrocera* species was studied by Gupta and Bhatia (2000) and found significant positive correlation with maximum and minimum temperature. Nath and Bhusan (2006a) recorded highest fly population in the 25th and 41st standard week during summer and rainy season respectively. However, they found positive correlation between relative humidity and fly population during summer and negative during rainy season. Positive and negative correlation was found with maximum and minimum temperature during summer and rainy season respectively. Rainfall showed positive and negative correlation with fly population in the summer and rainy season respectively. Nair (1995) observed that the flies congregate under leaves of plants in winter and become active when warm weather approaches. The peak population attained during rainy months of July and August. Singh and Naik (2006) found low incidence of the melon fruit fly, *Bactrocera cucurbitae* (Coq.) during January which increased gradually and attained its peak in March, thereafter declined subsequently. Among the weather factors, temperature favoured fruit fly damage while minimum relative humidity adversely affected pest damage.

Marked effects of temperature and rainfall on the population dynamics of the fruit flies, *B. dorsalis, B. cucurbitae* and *B. tau* was also noted by Deng *et al.* (2006). In a study conducted by Mahmood and Mishkatullah (2007) in Pakistan showed that population peak of the genus *Bactrocera* appeared in July and August and declining was observed in October depending on the host fruit maturity, temperature and rainfall.

Economic threshold level (ETL)

The economic threshold level (ETL) of melon fly is often based on the number of fruits infested in a given area as stated by Qureshi *et al.* (1976). Adult density of the fly responsible for causing 10% fruit infestation was recognized as economic threshold level (ETL) by Khan (1987). The fertilized adult female cause the damage by way of laying eggs under the skin of fruits and male adult flies are responsible for fertility of the female. So, the density of adult male flies is indirectly related with extend of infestation caused by the female as one male adult is supposed to represent at least one fertilized female or more than that.

The correlation co-efficient between fruit infestation and density of adult fly was found 0.97 by Inayatullah *et al.* (1991) in Pakistan. Simple linear regression model determined by them was Y=8.37+0.23x, where Y = percent fruit infestation and X= number of adult caught per trap per day that explained 95% variation (R^2=0.95) and concluded that for causing 10% fruit infestation, adult male density was 7/trap/day. The linear correlation between the number of puparia per unit (1X1ft. and 10cm.) soil sample was determined by the authors. They found the regression model as Y=3.10+9.31X, where Y= percent fruit infestation and X=number of puparia per unit soil sample that explained 85% variation. The authors also indicated through the model that 0.74 puparium/$ft.^2$ of soil resulted into 10% fruit infestation on squash and bitter gourd.

Management

Basically the cucurbits are summer season crops and most of them are grown well during warm months of the year. Pest problem is naturally high on these crops. Damage caused by the key pests is alarming specifically those inflicted by melon fly. Due to peculiar mode of infestation and multiplication of the pest it becomes hardly possible for the growers to combat it simply through pesticide application. Again, as the crops are vegetable, it is also troublesome for routine pesticide application on the crop against the pest complex. So, for sound pest management of cucurbits, critical life stages of the pests as well as site of occurrences to be targeted carefully.

The following details of methods used for detection, control and eradication are intended as a general introduction to the major techniques used:

i) **Detection**: Fruit flies may be detected during fruit import, in field by monitoring with traps or manual detection through sampling infested fruits. Acoustic detection system is now being developed to listen for larvae in individual fruits (Webb *et al.*, 1988). These techniques will help in restricting spread of fruit flies from one area to another.

ii) **Prevention**: Preventing fruit flies from becoming established in fruit fly free areas through strictly enforcing quarantine regulations. For example, restriction in the import of fruit fly susceptible crops from fruit fly prone areas without disinfestations treatment and forbidding travelers to carry fruit in their baggage.

In areas where labour cost is low, large fruits may be individually wrapped in paper or cloth before they reach suitable stage for fruit fly attack. Yeild of bitter gourd and angled luffa was increased by about 45% when the fruits were wrapped with two layers of paper bags, replaced every 2-3 days interval (Fang, 1989).

iii) **Collection and destruction of affected fruits:** Sanitation is one of the most important and elementary procedures in reducing favorability of the agro-ecosystem for the pest species because many species breed and overwinter in all sorts of debris and removal of debris from habitats can reduce rates of reproduction and survival. When an infestation is detected in the field it is important to gather all fallen and infested host fruits and destroy them. Burying of infested fruits at a depth from which the adults will not be able to come out. It should be not less than 50 cm. from the soil surface. Putting the infested fruits in plastic carry pack and placing them in sun shine will also be an effective way to destroy the maggots that remain within the infested fruits.

Collection and destruction of fallen, damaged, over-ripe and excess ripe fruits are strongly recommended to reduce resident population of fruit flies in all kind of fruit hosts. Srivastava (1996) suggested systematic and regular destruction of the affected fruits, constant stirring of the soil under the infested plants as the effective cultural management tactic of melon fruit fly management. Hasyim *et al.* (2008) opined that to break the reproduction cycle and population increase, growers need to remove all unharvested fruits or vegetables from a field by completely burying them deep into the soil. Klungness *et al.* (2005) reported that the most effective method in melon fruit fly management was by using field sanitation and also observed that burying damaged fruits 0.50 m deep into the soil prevent adult fly eclosion and thereby reduces population.

Vijaysegaran (1985) reported that the orchard sanitation by collecting and destructing all unwanted fruits on the trees and on the ground contribute significantly in reducing the fruit fly population. In Hawaii, *papaws* left on the ground act as a major breeding site for oriental fruit fly (*B. dorsalis*) and melon fruit fly (*B. cucurbitae*) and to eliminate or reduce this resident population reservoir, crop sanitation was suggested as an essential component of melon fly and oriental fruit fly management in *papaw* orchards in Hawaii. Yang (1991) in China noted that *B. citri*, a serious pest of citrus was successfully controlled by orchard sanitation.

iv) **Trapping**: Males of the most *Bactrocera* can be collected in traps which have been baited with special chemicals, sometimes called parapheromones. The most important male lure of especially of *B. cucurbitae* is Cuelure. In spite of *B. cucurbitae* it attracts some other Bactrocerans also. Its chemical description is 4-(*p*-acetoxyphenyl)-2-butanone. Very similar but less effective chemicals attracting the same range of species are anisylacetone and Willison's lure (Drew, 1982). It is also possible to 2-butanone, but it is highly volatile and must be kept in a slow release dispenser rather than being placed in cotton wick.

Traps that use male lures are usually based on the steiner trap design. It is a horizontal cylinder with a large opening at each end and the chemical lure impregnated into a cotton wick suspended within the trap. To avoid escaping of attracted and captured flies an insecticide is normally mixed with the lure or a strip of filter paper impregnated with insecticide is also placed within the trap which is also known as 'attracticide' as per Pedigo (1995). OP, DDVP are the preferred options of insecticide. Some other insecticides may alter the structure of the lure and render the bait ineffective. For monitoring and management of fruit flies, 10-12 traps/ha is recommended.

Both females and males of fruit flies may be collected in traps baited with a substance that emits ammonia fumes viz. yeast autolysate, hydrolysed protein or ammonium carbonate. The traps rely on the ammonia bait are usually based on McPhail trap. This type of trap has an entrance hole in its base and a trough around the base to hold a liquid bait such as a protein solution (Drew, 1982).

v) **Food bait:** Sounder Rajan *et al.* (1996) reported that moist fishmeal was more attractive to the tephritid than fermented palm juice, methyl eugenol, fermented mollasses, jack fruit and rice gruel. Banana was found to be as attractive as soybean hydrolysate probably because of its higher sugar content (Bose and Mitra, 1990) since fermenting sugars attracts fruit flies (McPhail, 1937). Lall and Singh (1969) used palm juice and dried mango juice along with citronella oil to trap adult *B. cucurbitae*. The flies respond positively to the sources of ethyl alcohol and acetic acid (Barrows, 1907). Usually the baits in traps dry up in 2-3 days due to weather factors (Taneja *et al.*, 1986) and once the moisture content decreases the fermentation process considerably reduced with little release of volatile compounds from the baits. Nath and Bhusan (2006) found that maximum number of adult fruit fly (*Bactrocera cucurbitae*) was trapped in the banana based poison bait trap containing banana (1kg) + carbofuran (10gm) + citric acid (5gm).

vi) **Insecticidal management**

- **Cover spray:** It is nothing but the conventional insecticide application on crop vegetation.
- **Bait spray:** It works on the principle that both male and female tephritids are strongly attracted to a protein source from which ammonia emanates. Bait sprays have the advantage over cover sprays that they can be applied as a spot treatment so that the

flies are attracted to the insecticide and there is minimal impact on natural enemies and residual toxicity in the produce. Both cover and bait sprays may be used for their *control* or *suppression*. The distinction between them being that, control refers to procedures designed to protect a single orchard while suppression refers to procedures covering a large area (Bateman, 1982). Malathion or Spinosad have shown good efficacy in managing fruit flies through bait spraying. In feeding trials, ingestion of spinosad resulted in high mortality for mexican fruit fly, *Anastrepha ludens* (Loew.) (Prokopy *et al*., 2000), caribbean fruit fly, *Anastrepha suspense* (Loew.) (King and Hennessey 1996) and mediterranean fruit fly, *Ceratitis capitata* (Wied.) (Adan *et al*., 1996). In field trials, Peck and McQuate (2000) demonstrated that bait sprays with malathion or spinosad suppressed *C. capitata* populations. However, malathion seemed to be more effective than spinosad.

- **Soil application of pesticides:** The mature maggots come in contact with soil for pupation and hence soil application of toxic chemicals have also found effective in minimizing fruit fly population. Dichrotophos (@ 600 gm a.i./ha) and trichlorphon (@ 1920 gm a.i./ha) has been found to give good control of *B. cucurbitae* in muskmelon (Chughtai and Baloch, 1988). However, Talpur *et al*. (1994) found formathion more effective than trichlorphon. Neem oil (1.2%) and neem cake (4.0%) have been reported to be as effective as Dichlorvos (0.2%) (Ranganath *et al*., 1997). Bio-efficacy and persistent toxicity of different insecticides including neem against melon fruit fly were tested by Sood and Sharma (2004) and found that maximum yield was obtained with deltamethrin treatment followed by cypermethrin, fenvalerate, malathion, deltamethrin + achook, deltamethrin + neemjeevan, achook and econeem. The synthetic pyrithroids had higher persistent toxicity (seven days) compared to malathion or the neem derivatives (three days or less).

vii) **Biological control**

Biological control of some fruit fly species has been tried including *B. cucurbitae* but the introduced parasitoids have had little impact in managing pest (Wharton, 1989).

During field tests in Hawaii conducted by Lindegren *et al*. (1990) it was revealed that drench application of entomopathogenic nematode *Steinernema feltiae* Filipjiv. to soil increased the mortality of mediterranean fruit fly larvae by 99.5%.

2.1.2 Pumpkin beetles: Red pumpkin beetle, *Aulacophora* (Raphidopalpa) *foveicollis* (Lucas), *Blue pumpkin beetle, A. lewisii* (Baly); Gray pumpkin beetle, *A. cincta* (Fab.) (Chrysomelidae: Coleoptera)

Hosts of commercial importance

The pests are basically polyphagous in nature though their preferred hosts are cucurbitaceous vegetables. In addition to cucurbitaceous vegetables they also found to infest some leguminous crops (Srivastava and Butani, 2009). The pest also found to infest Japanese mint, a cash crop of medicinal importance (Singh and Gupta, 1970), jowar and sorghum (Tayade and Chunderwar, 1977) and Lucerne (Chand and Singh, 1976) in India.

Distribution

The beetles have been reported damaging cucurbit crops. Among them the first one is the most damaging and widely distributed all over India. Also found to infest crops in Mediterranean region towards West and Australia in the East. In India, the pest is found in almost all the states. However, it is more abundant in northern states. *A. lewisii* is usually more common in northern tract of the country and *A. cincta* in southern tract. It is also reported to infest different cucurbits in Pakistan (Khan and Wasim, 2001).

Identification

Adults

A. foveicollis: The adult beetles are 6-8mm. in length, slender body, the elytra are glistening yellowish-red to yellowish-brown in colour and outer surface of the elytra are uniformly covered with fine punctures.

A. lewisii: Adult beetles are to some extend shorter in length as that of *A. foveicollis* which is about 5-6mm. Colour of elytra is blackish blue.

A. cincta: Adults of the species are more or less similar in size and appearance with the *A. foveicollis*. But the colour of elytra is grayish-yellow to brownish-green with distinct paler margin all around the wing.

Grubs and pupa: Freshly hatched grubs are dirty white in colour. On the other hand, the full grown grubs are creamy yellow. The full grown grubs are about 22 mm. long. Pupation takes place in soil in earthen cell at about a depth of 15-25mm from the soil surface and the pupae are pale white in colour.

Eggs: Eggs are spherical in shape and yellowish pink in colour and become orange colour after a couple of days after laying.

Biology

Eggs are laid by the adult female in moist soil usually around the host plant. Fecundity is about 150-300. The females do not prefer to lay eggs in dry and waterlogged soil. Incubation period is about 5-15 days depending upon environmental condition especially temperature. Larval, pre-pupal and pupal stage varies from 13-25, 2-5 and 7-17 days respectively. Total life cycle occupies about 32-65 days as noted by Narayanan (1953).

The adults live for about a month. In extreme cold temperature the overwintering takes place in adult stage most of which are males and about 75% males perish during the cold months of the year. The pest population in this period may comprise six females to one male (Srivastava and Butani, 2009).

Mode of feeding and symptoms of damage

Immediately after hatching, the grubs feed on the roots and underground portion of the plants as well as the fruits touching the soil. The damaged underground portion of the plants starts rotting due to secondary infection by several saphrophytic microorganisms. The immature fruits of such vines eventually dry up. The infested fruits also get rottened due to secondary infection and become unfit for human consumption.

The adult beetles feed voraciously on leaf lamina making irregular holes. They also feed on the underside of cotyledonary leaves by biting. The pest also reported to damage flowers, stems and even fruits (Johri and Johri, 2003). They prefer young seedlings and tender leaves and extreme damage at this stage the crop may need to be resown. In Pakistan the pest causes 35-75% damage to all cucurbits as reported by Saljoqi and Khan (2007). The authors also mentioned that the percent damage rating gradually decreases from 75-15% as the leaf canopy increases.

Seasonal Incidence

Maximum damage by the overwintering beetle during March to May. However, Butani (1975) found that the pest become active from March to October, though the peak period of activity is April to June. It was also found in a population study lasting for 12 months that the average sex ration was one male to 2 females but that the proportion of females declined from September onwards and rose from March onwards (Shinde and Purohit, 1978). The observation was directly correlated with the peak activity period when the preferred food plants are available. In Pakistan, infestation of the beetle has been found started in the beginning of May which subsequently increased and reached to its peak in the first week of June, while from June to August the population gradually declined Saljoqi and Khan (2007).

Management

a) To prevent damage by the pest, cultural practices like field sanitation, early planting and use of tolerant or resistant varieties have been recommended.

b) Immediately after harvest the field should be deep ploughed to kill the soil dwelling grubs directly or make them to be predated.

c) Panji (1964) stated that 10% ethanolic extract of dried fruit of *Melia azedarach* provide repellency effect on the beetle from cucurbitaceous vegetables.

d) The seed extract of *neem* showed very good repellent properties against the pest as reported by Chakravorty *et al.* (1969).

e) When the pest population reaches ETL insecticidal management is recommended. However, a good number of insecticide have shown phytotoxic effects on cucurbitaceous crops, great care should be taken in selecting proper insecticide. Makdoomi and Ishaq (1970) suggested carbaryl for the management of the pest that showed 90.68% mortality. Khan *et al.* (1985) reported that carbaryl dust mixed with ash in the ratio of 1:15 and dusted in the morning when there is dew drops on the plants is very much effective in managing the pest. Fenitrothion is also effective in managing the pest as reported by Bajwa and Mavi (1988).

f) Soil application with Furadan 3G @ 5g/plant at 3 days before planting, mechanical control with sweeping net at 3 days interval for 45 days, spraying neem seed oil @10ml/1+5m1 trix (detergent) at 7 days interval, spraying neem seed karnel extract @ 50g/1 of water at 7 days interval, seedling bed covered with mosquito net barrier upto 45 days old seedlings, controls red pumpkin beetle reported by Khorsheduzzaman *et al.* (2010).

g) Use of carbaryl, deltamethrin and dichlorvos as chemical and deep plough & winter cultivation of vegetables as non-chemical control method for the control of red pumpkin beetle (*Aulacophora foveicollis*Lucas.) reported by Saleem *et al.* (2010).

h) Mixing of carbaryl 10 % WP in pits before sowing the seeds destroys grubs and pupae. Spraying of Sevin 50 WP (2 g/L) was also effective control measure reported by Bharati *et al.* (2013).

i) Application of combination of Carbaryl as dust + NSKE as spray + Yellow sticky trap as a mechanical control was the most appropriate for red pumpkin beetle management reported by Rashid *et al.* (2015).

2.1.3 Hadda beetle: *Henosepilachna vigintioctopunctata* (Fab.), *H. dodecastigmata* (Coccinellidae: Coleoptera)

Hosts of commercial importance:

The pests are highly polyphagous in nature and cause damage to agricultural crops in three families *viz.* Solanaceae (potato, tomato, aubergine and pepper), Cucurbitaceae (cucumber, melon, water-melon and pumpkin) and Fabaceae (soya and haricot beans).

It is the major pest of several vegeta bles including eggplant, cucurbits, potato and peas, tomato and related plants (Olliff, 1980), tobacco, *Datura stramonium,* cotton (Froggatt, 1923) and also medicinal plant 'ashwagandha' (Manjoo and Swaminathan, 2007; Chandranath and Katti, 2010). Among cucurbitaceous vegetables, bitter gourd is the preferred one (Srivastava and Butani, 2009).

Distribution

This species originated in the far east of Russia and has been expanding its range in the second half of the 20th century and is now found all over Russia and other parts of the world. Among the major insect pests that attack cucurbits, the Epilachna beetle, *E. vigintioctopunctata* is very important in Asia. Their range extends throughout India, Pakistan, China, Japan, SE Asia and Oceania. Nine species *Epilachna* have also been recorded from Australia (Richards, 1983). It is widely distributed in South and East Asia, Australia, America, and the East Indies (Rajagopal and Trivedi, 2008). These two epilachna beetles are commonly found in India. However, these are seen in two interbreed and may be considered as only variations of a single species (Nair, 1995).

Identification

Adults: Beetles are about 8-9mm in length and 5-6mm in width. *E. vigintioctopunctata* beetle is nearly round, convex, glossy, deep red and usually have 7-14 black spots on each elytron and one or more on each side of the thorax, tip of the abdomen is pointed. Beetles of *H. dodecastigmata* are deep copper coloured and six black spots on each elytron. Tip of the abdomen in this case is to some extend rounded. The mean lengths and widths of male and female adults were 5.85±0.32, 6.85±0.23 mm and 4.55±0.42, 5.05±0.15 mm respectively as noted by Ramandeep and Mavi (2005).

Grubs: Grubs of all the species are oval, about 6 mm in length and yellowish in colour. They have six rows of long branched thorny appendices. The length and breadth of final instar grubs were highest feeding on leaves of sponge gourd (7.90±0.27 mm, 6.95±0.43 mm) which were statistically identical to teasel gourd (7.82±0.38 mm, 6.93±0.09 mm) and bitter gourd (7.71±0.09 mm, 6.83±0.38 mm). The lowest length and breadth of grub were recorded when fed on yard long bean (5.32±0.22 mm, 3.20±0.13 mm) (Hossain *et al.*, 2009).

Pupa: The pupae are hemispherical with mean length and width of 7.05±0.44 and 4.00±0.41 mm respectively.

Eggs: The eggs are yellow, cigar-shaped, about 1.5 millimetres long. The length and breadth of egg were highest in teasel gourd (1.09±0.04 and 0.38±0.05 mm) which was statistically identical to those of sponge gourd and bitter gourd and the lowest was in yardlong bean (0.60±0.12 and 0.35±0.02 mm) (Hossain *et al.*, 2009).

Biology

Eggs are placed by the female on undersides of leaves perpendicularly with the leaf surface in batches close to one another of ten to sixty five eggs. Incubation period lasts for 3-4 days depending upon environmental conditions. Fecundity is about 120-180 as noted by Nair (1995). However, Hossain *et al.* (2009) observed highest fecundity on teasel gourd leaves (268.00±33.99) followed by sponge gourd, bitter gourd, and yard long bean. The author also found longest incubation period on bitter gourd (4.19±0.23 days) and shortest in yard long bean (3.15±0.32 days). Ramandeep and Mavi (2005) recorded the number of eggs laid by a single female as 114.80±52.85. In south India (Karnataka) the larval period lasts for 13-15 days on potato and 30-32 days on cucurbits during February to May (Nair, 1995). Average incubation and larval period were found shortest on brinjal and longest on *datura* (*Datura stramonium*) (Dhamdhere *et al.*, 1990). In Bangladesh, larval duration on sponge gourd was observed 11.38±0.53 days and on bitter gourd 11.05±0.82 days. Teasel gourd showed significantly lowest larval duration of 9.58±0.33 days. Pupation takes place on infested parts on the plant and pupal period varies from 3-6 days depending upon environmental factors. Longevity of males and females recorded on different food plants ranged from 29 to 43, and 34 to 57 days respectively.

Impact of different host plants on growth and development

When epilachna beetle was reared in the laboratory, potato was found to be the most suitable host and the coccinellid beetle completed its life cycle very quickly on *Solanum nigrum* in 22.4 days. Chen *et al.* (1989) studied the biology of this coccinellid and observed that a single generation lasted for 23.1 to 35.7 days at varied level of temperature. In an another laboratory study conducted by Srivastava *et al.* (1969) with five solanaceous plants (tomato, brinjal, *S. insanum*, *Datura stramonium* and *Physalis maxima*) for comparison with respect to their suitability as food plants and it was found that at 25-29°C and 88-94.5% R. H. the average duration of the larval period ranged from 13 days on *P. maxima* to 16.35 days on brinjal. Parjhar *et al.* (1997) tested six solanaceous plants for suitability as food plants for *Epilachna vigintioctopunctata* Fab. and found brinjal and potato the best with regard to survival and duration of larval development. The other plants tested were *Solanum nigrum*, *Nicandra physaloides*, *Withania somnifera* and *Lycopersicon esculentum*.

Nagia *et al.* (1992) found larval and pupal periods of *H. vigintioctopunctata* as 11.54±0.69 and 3.57±0.50 days respectively on *Physalis minima* and 14.20±0.65 and 4.65±0.49 days on brinjal. Females laid 194.60±20.99 and 147.40±23.48 eggs each on *Physalis* and brinjal. The average number of days required to complete the entire larval stage of this beetle was 15.95±0.86 and 15.60±0.73 days on tomato and brinjal respectively as revealed by Patel and Purohit (2000). They also found that adult longevity of the coccinellid was 23.40±0.65 and 27.55±0.51 days on tomato and brinjal respectively. Ghosh and Senapati (2001) recorded life cycle duration of epilachna beetle from 26.74 (June-July) to 33.52 days (September-October) and highest fecundity 272.32 (Eggs laid per female) during March to April under terai region of West Bengal, India. The gravid females of the beetle laid 287±33.38 eggs on an average during their oviposition period of 10.40±2.80 days as recorded by Venkatesha (2006).

The biology of *Epilachna vigintioctopunctata* Fab. was also studied by Ramandeep and Mavi (2005) and recorded the duration of pre-oviposition,

oviposition, post-oviposition and larval period period as 8.60±0.80, 11.80±1.54 and 6.90±0.83 days respectively. Four successive larval instars lasted for 2.20±0.40, 3.60±0.66, 5.70±0.46 and 4.10±0.54 days respectively. However, Araujo *et al.* (2004) recorded average duration of 1st, 2nd, 3rd and 4th instar developmental stages of *Epilachna vigintioctopunctata* as 5.88, 4.62, 5.88 and 9.81 days respectively when reared on tomato. In their study incubation period and egg viability was found as 7.14 days and 63.68% respectively.

Mode of feeding and symptoms of damage

Both grubs and adults are injurious to the host plants. Infestation primarily begins in colonial form just after hatching of egg mass (Murata *et al.*, 1994). The adults feed irregularly upon the upper surface of leaves and its grubs feed on the lower surface of leaves by scraping, causing net like appearance of the host plant leaves that turn brown in colour, entirely dry up due to extensive infestation by the growing population and finally defoliate (Pradhan *et al.*, 1990). Adults are active fliers and move about from plant to plant. On the other hand, the grubs are persistent on the leaves and occur in large numbers. Feeding of the grubs results in characteristic development of a symptom called 'windows' in which both feeding and non feeding portion remain parallel to each other.

Vegetative growth and development of the plants are greatly harmed and their economic yield is noticeably declined (Alam, 1969; Rajagopal and Trivedi, 1989). From economic stand point, the Epilachna beetle bears great significance because it can alone damage up to 80% of the host plants (Rajagopal and Trivedi, 1989), while it is responsible for 10-20% yield loss in brinjal (Alam, 1969).

Seasonal abundance

Insect pest prevalence and its degree of infestation vary from season to season, place to place due to the environmental changes especially temperature, and host plant species (Konar *et al.*, 2002). It is well recognized that the peak infestation by this insect pest is generally observed in July to August i.e. during the rainy season and usually there is no movement, activity or infestation by the pest found in winter season when they go to hibernation (Shukla and Upadhyay, 1985).

Epilachna beetles feed most actively during morning and evening hours. The daily fluctuation in the rate of feeding depends mainly on the temperature of the environment which estimates the level of metabolism (Tilavov, 1981). Tripathi and Misra (1991) observed the activity of coccinellid, *Epilachna dodecastigma* (Wied.) in a study conducted in the field in North-Eastern Uttar Pradesh, India and found that the populations were high during the period from late July to October and low in December and January. Hossain *et al.* (2009) found the peak period of infestation during July and August in Bangladesh.

Field studies were carried out by Ramzan *et al.* (1990) at Ludhiana, India on the developmental behaviour and seasonal abundance of epilachna beetle on various solanaceous food plants showed highest population of 526.3/ 10 plants in March. Venkatesha (2006) studied seasonal occurrence of the beetle on a medicinal plant, *Withania somnifera* in Bangalore (Karnataka, India) during 2004-05 and noticed peak population level in August.

Management

a) Collection and destruction of egg masses as well as adults and grubs is effective maintaining regular close vigil in the field particularly small scale cultivation.

b) A number of different groups of insecticides have also been evaluated against the pest by many workers (Wei *et al.*, 2004 and Singh *et al.*, 2005) all over the world in various hosts and many of them found very much effective in controlling it.

c) Sahu *et al.* (1980) studied the relative toxicity of five insecticides in comparison with carbaryl against 3rd instar grubs of *Epilachna sparsa* and found that fipronil, diflubenzuron, triazophos, thiodicarb and cartap hydrochloride were 1.42, 0.69, 0.58, 0.31 and 0.08 times toxic by leaf dip method and 1.54, 14.73, 1.51, 0.64 and 0.34 times toxic by direct spray method as compared to carbaryl. Except for diflubenzuron, the test compounds exhibited 1.87 to 7.94 times higher toxicity in leaf dip method than that in direct spray method.

d) Pareek and Kavadia (1988) determined the effectiveness of several insecticides against epilachna beetle on muskmelon and found all the insecticides significantly reduced the pest population as compared with untreated plots. They also observed that carbaryl at 0.2% was the most effective followed by phosalone at 0.03%.

e) Nine insecticides were evaluated in the laboratory at 26 C and 70% R.H. and L:D=12:12 against larvae and adults of *Henosepilachna vigintioctopunctata* (Fab.) on aubergine by Nagia *et al.* (1992). The findings revealed that the carbaryl, quinalphos, cypermethrin, deltamethrin and fenvalerate were effective against larvae and adults while dimethoate, fenpropathrin and fluvalinate were effective against larvae only.

f) Konar *et al.* (2005) evaluated bioefficacy of some insecticides and biopesticides against *Henosepilachna vigintioctopunctata* (Fab.) on potato. The treatments comprised 1.0 kg carbofuran /ha, 0.5kg chlorpyriphos /ha, 1.0 kg. carbaryl /ha, 1.0 kg acephate /ha, 1.0 kg *Beauveria bassiana* /ha (at 108 spores/ml.) and 1 litre azadirachtin / ha (at 10000 ppm/ha).

   Due to good efficacy of several insecticides most of the farmers are using insecticides indiscriminately for control of this notorious pest. But indiscriminate use of insecticide has not only complicated the management but also created several adverse effects such as pest resistance, outbreak of secondary pests, health hazards and environmental pollution (Fishwick, 1988).

g) 5% of both aqueous and petroleum ether extract of *Calotropis gigantia* leaves (Konimozhi and Veeravel, 2007), *Annona squamosa, Argemone mexicana, Cacopris gigantea* and *Ricinus communis* leaves (Rao *et al.*, 1990) were found to be most promising in protecting the leaf damage due to attack of the pest.

h) The repellent properties of neem, mahua and groundnut have been established by Rajagopal and Trivedi (1989). The authors also observed that *Aspergillus flavus* and *Bacillus thuringiensis* are effective against different stages of the pest.

i) Chitra *et al.* (1992) also tested 18 different plant extracts @ 0.5 and 1.0% against the pest which provided above 90% protection of the crop by a number of plant extracts. They also noted over 20% reduction in the weight of larvae.

j) Islam *et al.* (2011) observed that crude aqueous extracts of leaves *Ricinus communis* had highest larvicidal toxicity (LC_{50}=18.40%) besides significant reduction in both oviposition and egg-hatch, prolonged larval duration ($P<0.001$), and inhibited pupae formation and adult emergence.

k) Need based application of malathion or carbaryl @ 1 kg a.i. /ha or cypermethrin @ 0.4 ml/l is able to control this polyphagous pest as reported by Rai *et al.* (2014).

l) Vishwakarma *et al.* (2011) documented that, significantly maximum reduction in the population of Epilachna beetle (74.91%) was achieved in treatment of *Beauveria bassiana*, when used @ 3.0 g/l of water.

2.1.4 Pumpkin caterpillar: *Diaphania indica* (Saunders) (Pyralidae: Lepidoptera)

Diaphania indica, Cucumber moth is one of the serious pests of plant under cucubitaceae. However, Srivastava and Butani (2009) found that the pest is a minor one that found on almost all cucurbitaceous crops. It is also known as cotton caterpillar.

Hosts of commercial importance

It infests cucumber, melon, gherkin, bottle gourd, bitter gourd, snake gourd, Luffa, little cucumber, cotton etc. It is also reported to infest crops of other families viz., Leguminosae and Malvaceae. Field and laboratory studies by Ba-Angood (2009) have shown that the insect prefers *Cucumis melo* (sweet melon) than *Citrulus vulgaris* (water melon) and *Cucumis sativa* (cucumber).

Distribution

It is distributed in India, China, Japan, South Korea, Bangladesh, Saudi Arabia, Yemen, sub-Saharan Africa, USA (Florida) and Northern Territory of Australia.

Identification of the pest

Caterpillars are elongated in shape and bright green in colour having two narrow longitudinal stripes dorsally. Adults are medium sized moths, have translucent whitish wings with broad dark brown borders, wingspan is about 30 mm. The body is whitish below, and brown on top of head and thorax as well as the end of the abdomen. There is a tuft of light brown to orange coloured "hairs" on the tip of the abdomen, vestigial in the male but well-developed in the female. It is formed by long scales which are carried in a pocket on each side of the 7th abdominal segment, from where they can be everted to form the tufts. Unfertilized females are often seen sitting around with the tuft fully spread, forming two flower-like clumps of scales, which move slowly to spread their pheromones.

Biology

A female laidon an average of 187.1 eggs on leaves of the host plants. On hatching, the young larvae lacerate and feed on chlorophyll of foliage. The average egg, larval, prepupal and pupal stages lasted for 4.75, 11.9, 1.3 and 9.4 days, respectively. The female to male sex ratio is about 1.08:1. The longevity of adults varied from 8.45 to 9.0 days. The mean developmental period from oviposition to adult emergence ranged from 23 to 33 days with an average of 27.35 days. Shashpa (2004) reported that the pre-oviposition, oviposition and post-oviposition periods of *Diaphania indica* averaged 2.0, 5.1 and 1.5 days, respectively. Incubation period, larval period, pupal period lasts for 4-6, 11-14 and 5-15 days respectively as mentioned by Srivastava and Butani (2009). The effect of different constant temperatures on the development of the insect has shown that as the temperature increases, egg incubation period, larval and pupal periods decreases considerably. At fluctuating temperatures of 25–35.9 °C, the egg, larval and pupal periods were 3.5, 11.5 and 6.2 d, respectively (Ba-Angood, 2009).

Mode of feeding and symptoms of damage

Eggs of *D. indica* are singly laid in small clusters on the lower surface of leaves. On hatching, the larvae feed on the lower surface of leaves which are lacerated and bound together by threads of silk. In severe outbreaks, most of the foliage get destroyed and larvae burrow into the stems. The fruits are sometimes attacked particularly at the proximal end. Caterpillars have also been observed to damaging the ovaries of flowers and boring into young developing fruits (May, 1946).

The young larvae fed on chlorophyll and skeletonizes the leaves of cucurbits, while the older larvae folded and webbed together the leaves and fed within. The young and older larvae were also observed feeding on flowers and boring into ovaries, new tender shoots and young and developing as well as mature fruits in the case of little gourd and bitter gourd. Observations showed that the damage to the fruit was 90% in little gourd and 60% in bitter gourd. In the case of pointed gourd, the damage by the larvae was restricted to leaves only and it was 25-30%. Common mynah (*Acridotheres tristis*) and cattle egret (*Bubulcus ibis*) were found predating on larvae of *Diaphania Indica* (Jhala *et al.*, 2004).

Seasonality of occurrence

Incidence of the pest may be attributed to changes in cropping pattern and geographical location and environmental conditions. Peak period of infestation was also noted at maximum flowering stage of the crop at 89 DAS in cucumber (Vanisree *et al.*, 2005). The authors also detected that morning relative humidity is the most important abiotic component of the environment that influence incidence of pumpkin caterpillar.

Management

a) In early stages of infestation, handpicking of caterpillars and their mechanical destruction helps in keeping the pest population under check.

b) Moth catcher using pheromone may be used in dusty condition or in high moth population density. Decisions on pesticide application should not be taken solely on the trap catch data. Climatic and biological considerations should be taken into account.

c) Use of plant-derived products, such as neem, derris, pyrethrum and chilli (with the addition of soap), or commercial products that contain disease-causing organisms, such as spinosad (Success) and Bt - *Bacillus thuringiensis* subspecies *kurstaki*. (NIPHM, AESA BASED IPM Package No. 21, 2014) are effective.

d) Pyrethrum and derris insecticides can also be used; these, too, are made from natural products. They are fast acting against insects and less harmful to the environment compared to many synthetic (commercial) products. They are broken down quite quickly by sunlight. However, they will also kill natural enemies of Diaphania. (NIPHM, AESA BASED IPM Package No. 21, 2014).

2.1.5 Snake gourd semilooper, *Anavedidia* (*Plusia*) *peponis* Fab. (Noctuidae: Lepidoptera)

Hosts of commercial importance

This is a serious pest of snake gourd all over India (Nair, 1995). It is a defoliator of cultivated plants especially Cucurbitaceous crops viz. cucumbers and bottle gourds in Japan (Inomata, 2000).

Identification

The adult is a dark-brown stout moth with wing expanse of 30-35mm. Eggs are spherical in shape, beautifully sculptured and greenish white in colour and laid singly on the tender leaves. It grows to a length of about 30-40mm in length with whitish green elongated body having black darts, off white longitudinal stripes and is humped on its anal segment.

Biology

Egg, larval and pupal periods last for 4-5, 24-30 and 7-8 days respectively and life cycle is completed in about 6 weeks. Pupation takes place in leaf fold within a thin cocoon.

Mode of feeding and symptoms of damage

Caterpillar cut edges of leaf lamina, folds it over and hides within the leaf fold and feeds from within. The damage is often serious and when young plants are attacked they are totally denuded of their leaves. Infestation is thus become quite conspicuous.

Management

a) Collection and destruction of the caterpillars and application of contact poisons are the recommended practice for their management.

b) The caterpillar is parasitized by *Apanteles paragame* Vier., *A. plusiae* Vier. and *Mesochorus plusiaephilus* Vier.

c) Dusting Carbaryl 5% is effective for the management of this pest when population reaches or crosses ETL.

d) Spraying of malathion 50 EC @ 500 ml, dimethoate 30 EC 500 ml, methyl demeton 25 EC @ 500 ml/ ha was found to be effective (agritech.tnau.ac.in, 2015).

2.1.6 Bottle gourd plume moth: *Sphenarches caffer* (Zeller) (Pterophoridae: Lepidoptera)

Hosts of commercial importance

It is a polyphagous pest found to cause damage to pigeon pea, peas, cocoa, lablab, *Cucurbita pepo* and other cucurbits. Among the cucurbits, it has been reported damaging bottle gourd and *Luffa* sp.

Distribution

The pest is widely distributed in West, South and East Africa, Maldive Islands, Indian sub-continent, Phillipines, Japan, Indonesia, Australia and Tonga Islands in Oceania and West Indies (Fletcher, 1920).

Identification

Eggs are oval in shape, bluish green in colour with reticulate designs. Caterpillars are small, cylindrical, 7-9 mm. long when full grown, green in colour, having a lateral brown stripe on either side and clothed with dense pubescence of short spines and long capitate hair.

The adult is a tiny moth; wing span is about 10 mm. (0.39 in). Both the fore wings and hind wings are composed of feather-like plumes. These are buff with brown marks. When resting, the moth sits with its abdomen curved up into the air and its wings held at right angles to the body with the plumes folded.

Mode of feeding and symptoms of damage

The larva feed on the foliage and is a minor pest. Adult moths do not cause any direct damage to the crop.

Management

a) *Apanteles ruidus* Wlk. and *A. paludicolae* Cam. are the larval parasites of the pest. Any contact and stomach poisons are effective in managing the pest.

b) Need based application of *Bacillus thuringiensis* @ 1 kg/ha or malathion @ 2 ml/lit is able to control this pest as reported by Halder *et al.* (2014).

c) Collection and destruction of larvae and pupae is the effective management practice of the pest.

d) Spraying of malathion 50 EC @ 500 ml, dimethoate 30 EC 500 ml, methyl demeton 25 EC @ 500 ml/ha was found to be effective (agritech.tnau.ac.in, 2015).

2.1.7 Stem gall fly, *Lasioptera falcata* Felt. (Cecidomyiidae: Diptera)

Two species of gall midges have been reported on *Momordica charantia* Linn., namely *Lasioptera falcata* Felt. and an unidentified species.

Commercial hosts

Bitter gourd, squash and vegetable marrow.

Distribution

L. falcata was originally described by Felt (1919) on a single female reared from a stem gall on an unidentified wild cucurbit in the Philippines. It has been found to breed on *M. charantia* in South India. Ramakrishna (1920) mentioned these midges as the bitter gourd vine gall fly. Mani (1934) stated that it was very common at Coimbatore and had been found in various other localities in South India. In Bangladesh it is commonly known as gall gnats or gall midges, viz, *Lasioptera falcata,* and the larvae are seen in the stem of the bitter gourd.

Identification

Minute, delicate Dipteran insect characterized by beaded, somewhat hairy antennae and few veins in the short-haired wings. The brightly orange-coloured larvae usually causing the formation of tissue swellings (galls) in the distal tender shoots and feed from within. The maggots are free living, the head-capsule is well-developed, and mouthparts are of biting type. Adults are slender, dark brown and mosquito-like insect. They are identified from other flies by the presence of relatively long hair-like antennae, and long legs. The antennae are particularly well-developed in males. Male gnats fly in swarms, usually toward evening, dancing up and down a few feet from the ground.

Biology

Eggs are laid by adult female on tender shoots. Maggots bore inside the distal shoots and feed from within. Pupation also takes place within the galls. The infested vines develop galls. The galls are attennate-fusiform, solid indehiscent angulated and costate. But generally cylindrical swelling of branches, finally pubescent and otherwise smooth (Mani, 1973).

Mode of feeding and symptoms of damage

The larvae usually causing the formation of tissue swellings (galls) in the distal tender shoots and feed from within. The galls, which are long and tubular, are to be found on the distal young shoots. They affect the growth of the plant. Another species, *Dasyneura beccariella* Del Guercio, has been found on *Momordica pterocarpa* Hochst. *(pteromorfa)* in Eritrea. Gall formation causes stunting of the plants.

Galls are formed by feeding or egg-laying activity of the insect. Either mechanical damage or salivary secretions (introduced by insects) initiate increased production of normal plant growth hormones. These plant hormones cause localized plant growth that can result in increasing in cell size (hypertrophy) and/or cell number (hyperplasia). The outcome is an abnormal affected shoot structure called a gall. Gall formation generally occurs during the accelerated growth period of new shoots. Mature plant tissues are usually remain unaffected by gall-inducing organisms. Larvae of the gall-making insect develop inside the gall and the gall continues to grow as the insect feeds and matures. Once gall formation is initiated, many galls will continue to form even if the insect dies or undergo pupation. In addition, most galls are usually not noticed until they are fully formed and remain on plants for extended periods of time.

Galls are growing plant parts and require nutrients just like other plant parts. Its possible that galls steal vital plant food and adversely affect plant growth. This is most likely a problem when galls are numerous on very young plants. Injury may also occur if galls are numerous on branches or if abundant for several consecutive years. In most cases, however, galls are not abundant enough to harm the plant.

Management

a) Removal and destruction of affected shoots in early stage of infestation is the only effective management practice of the pest.

b) Chemical applications are an option, but are often ineffective since the precise timing of spray is critical. To be effective, sprays must be timed to coincide with initial insect activity before gall formation begins.

c) Once galls start to form, they conceal the causal organism and it is too late for treatment. For insects that overwinter on the host plant, horticultural mineral oil (HMO) applications can be made before insect activity begins.

d) Spraying of malathion 50 EC @ 500 ml, dimethoate 30 EC 500 ml, methyl demeton 25 EC @ 500 ml/ha was found to be effective (agritech.tnau.ac.in, 2015).

2.1.8 Serpentine leaf miner, *Liriomyza trifolii* (Burgess): (Agromyzidae: Diptera)

The serpentine leafminer, *Liriomyza irifolii* (Burgess) was first observed in Hawaii in 1978 after chrysanthemum growers failed to achieve economic control of leafminers with common insecticides such as aldicarb and methomyl (Nakahara 1982). There are 376 species currently recognized in the genus *Liriomyza* with 136 of these species found naturally in Europe (Seymour, 1994).

Hosts of commercial importance

It is an invasive pest and attacks a large number of plants, but seems to favour those in the plant families Cucurbitaceae, Leguminosae, and Solanaceae. Stegmaier (1966) reported nearly 40 hosts from 10 plant families in Florida. In India, it was initially recorded on 55 plant species (Viraktamath *et al.*, 1993) and later on about 79 species (Srinivasan *et al.*, 1995) that included pulses, oil seeds, vegetables, green manures, fodder and fibre crops. Galande *et al.* (2004) recorded this pest on 16 new crops and 16 weed species. Cowpea is one of the important legume crops affected by this pest.

Among the numerous weeds infested, *Solanum americanum* and *Bidens alba* are especially suitable hosts in Florida (Schuster *et al.*, 1991). Vegetable crops known as hosts include bean, eggplant, pepper, potato, squash, tomato, and watermelon. Celery is also reported to be attacked by the pest. In Hawaii, damage to onion foliage is a problem for the marketing of scallions (green onions) (Kawate and Coughlin, 1995).

Distribution

The pest is native to eastern United States and Canada, northern South America, and the Caribbean. However, in recent years it has been introduced into California, Europe, and elsewhere. Expanded traffic in flower crops appears to be the basis for the expanding range of this species. As a vegetable pest, however, its occurrence is limited principally to tropical and subtropical regions. The pest was accidentally introduced into India from American sub continent along with chrysanthemum cuttings (Anonymous, 1991a).

Identification

Adults: Adults are small, measuring less than 2 mm in length, with a wing length of 1.25-1.9 mm. The head is yellow with red eyes. The thorax and abdomen are mostly gray and black whereas, ventral surface of the body and legs are yellow. The wings are more or less transparent. The adults have grayish black mesonotum (dorsal portion of mesothorax), yellow hind margins of the eyes and yellow femora. Females are usually larger in size and more robust than the males, and have an elongated abdomen.

Flies normally live only about a month. Population is meagre during the cool months of the year, but often attains high, damaging levels by mid-summer. In warm climates they may breed continuously, with many overlapping generations per year.

Eggs: The sexually mature female deposits the eggs on the lower surface of the leaf and are inserted just below the epidermis. Eggs are oval in shape and small in size, measuring about 1.0 mm long and 0.2 mm wide. Initially they are clear but soon become creamy white in color as they come close to hatching.

Larvae: A full grown larva attains a length of about 2.25 mm. and passes three active instars. Initially the larvae are nearly colorless, becoming greenish and then yellowish as they mature. Mouthparts are apparently black in all instars. The average length and range of the mouthparts (Cephalopharyngeal skeleton) in the three larval feeding instars is 0.09 (0.6-0.11), 0.15 (0.12-0.17), and 0.23 (0.19-0.25) mm, respectively. The mature larva cuts a semicircular slit in the mined leaf just prior to formation of the puparium. Almost invariably, the slit is cut in the upper surface of the leaf. The larva usually emerges from the mine, drops from the leaf, and burrows into the soil to a depth of only a few cm to form a puparium. A fourth larval instar occurs between puparium formation and pupation, but this is generally ignored by the authors (Parrella, 1987).

Pupa: The reddish brown puparium measures about 1.5 mm in length and 0.75 mm in width. After about nine days of pupation, the adult emerges out from the puparium, principally in the early morning hours, and both sexes emerge simultaneously.

Biology

Mating of sexually mature adults initially occur during the day following adult emergence. Eggs are inserted into plant tissue just beneath the leaf surface and hatch in about three days. Flies feed on the plant secretions caused by oviposition, and also on natural exudates of feeding punctures, particularly along the margins or tips of leaves, without depositing eggs. Leibee (1984), working with celery as a host plant, estimated that oviposition occurred at a rate of 35 to 39 eggs per day, for a total fecundity of 200 to 400 eggs. In India, Jeyakumar (1995) reported that the fecundity varied with hosts and it was 113.1, 142.2 and 130.0 eggs on cotton, tomato and cowpea, respectively. Initially, females may deposit eggs at a rate of 30 to 40 per day, but egg deposition decreases as flies grow older. Development rates increase with temperature up to about 30°C; temperatures above 30°C are usually unfavorable and larvae experience high mortality.

Minkenberg (1988) indicated that at 25°C the egg stage required 2.7 days for development; the three active larval instars required 1.4, 1.4, and 1.8 days, respectively; and the time spent in the puparium was 9.3 days. Also, there was an adult preovipostion period that averaged 1.3 days. The temperature threshold for development of the various stages is 6 to 10°C except that egg laying requires about 12°C. Duration of different life stages varies depending upon prevailing environmental condition. Thus incubation period varied from 2.0 to 8.0 days, larval period also varied from 4 to 25 days (Ganapathy, 2010). Under Indian conditions, adults lived for 3 to 6 days (Lakshminarayana *et al.*, 1992) and male and female longevity varied from 4.7 to 6.0 days in cowpea (Jeyakumar, 1995).

Mode of feeding and symptoms of damage

Punctures caused by females during the feeding and oviposition processes can result in a stippled appearance on foliage, especially at the leaf tip and along the leaf margins (Parrella *et al.*, 1985). However, the major form of damage is the mining of leaves by larvae, which results in destruction of leaf mesophyll. The mine becomes noticeable about three to four days after oviposition and becomes larger in size as the larva matures. The pattern of mining is irregular. Both leaf mining and stippling can greatly depress the level of photosynthesis in the plant. Extensive mining also causes premature leaf drop, which can result in lack of shading and sun scalding of fruit. Wounding of the foliage also allows entry of bacterial and fungal diseases. Although leaf mining can reduce plant growth, crops such as tomato are quite resilient, and capable of withstanding considerable leaf damage.

It is often necessary to have an average of one to three mines per tomato leaf before yield reductions occur (1975, Schuster *et al.*, 1976). In potatoes, Wolfenbarger (1966) reported sizable yield increases with as little as 25% reduction in leafminer numbers, but leafminers were considered to be less damaging than green peach aphid, *Myzus persicae* (Sulzer).

Management

Leaf miner assessment

Counting mines in leaves is a good index of past activity, but many mines may be vacant. Counting live larvae in mines is time consuming, but more indicative of future damage.

Puparia can be collected by placing trays beneath foliage to capture larvae as they evacuate mines, and the captures are highly correlated with the number of active miners.

Adults can be captured by using adhesive applied to yellow cards or stakes. Sequential sampling plans for determining the need for treatment, based on the number of mined leaves were developed by Wolfenbarger and Wolfenbarger (1966).

a) Cultural Practices:

Because broad leaved weeds and senescent crops may serve as sources of inoculum, destruction of weeds and deep ploughing of crop residues are recommended.

Adults experience difficulty in emerging if they are buried deeply in soil. Cultural practices such as mulching and staking of vegetables may influence both leafminers and their natural enemies.

Price and Poe (1976) reported that leafminer numbers were higher when tomatoes were grown with plastic mulch or tied to stakes. At least part of the reason seems to be due to lower parasitoid activity in plots where tomatoes were staked.

b) Biological Control:

As the parasitoids often provide effective suppression of leafminers in the field when disruptive insecticides are not used, there has been interest in release of parasitoids into crops.

This occurs principally in greenhouse-grown crops, but is also applicable to field conditions. Parrella *et al.* (1989) published information on mass rearing *Diglyphus begini* for inoculative release.

Steinernema nematodes have also been evaluated for suppression of leaf mining activity. One of the most complete studies, by Hara *et al.* (1993), showed that high levels of relative humidity (at least 92%) were needed to attain even moderately high (greater than 65%) levels of parasitism.

c) Insecticides:

Chemical insecticides are commonly used to protect foliage from injury, but insecticide resistance is also a major problem. Insecticide susceptibility varies widely among populations, and level of susceptibility is directly related to frequency of insecticide application.

Insecticides are highly disruptive to naturally occurring biological control agents, particularly parasitoids. Use of many chemical insecticides influence leafminer problems by killing parasitoids of leafminers. This usually results when insecticides are applied for lepidopterous insects. Use of more selective pest control materials such as *Bacillus thuringiensis* is recommended as it allows survival of the leaf miner parasitoids.

Efficacy of neem seed kernel extract, neem oil, *karanj* oil and *mahua* oil was recorded by Stein and Parrella (1987) and Azam (1991). The effectiveness of chlorpyriphos and triazophos against serpentine leaf miner was documented by Jeyakumar (1995) on cotton.

In addition to these,

i) Application of NSKE 4% with a sticker found to be effective to deter the leaf miner. In case of severe infestation, spraying of

Imidacloprid 17.8SL @ 0.3 ml/l of water during early stages of crop growth before flowering is effective as reported by Nadagouda *et al.* (2010).

ii) Kumar *et al.* (2010) reported that seed treatment with Imidacloprid 70WS @ 3g/kg seeds or foliar applications of neem seed kernel extract (NSKE) 5% on 10 days after germination gives better result in controlling the serpentine leaf miner.

iii) Rai *et al.* (2014) mentioned that spraying of Imidacloprid 17.8 SL @ 0.35 ml/l of water during the early stage of the crops before flowering and application of Dichlorovos 76EC @ 0.5 ml/l of water in severe infestation during reproductive phase crop is beneficial.

2.1.9 Stem borer or clear winged moth: *Melittia cucurbitae* Westwood (Sesiidae: Lepidoptera)

Hosts of commercial importance

Its preferred vine crops are winter squash, summer squash and pumpkins and less frequently on cucumbers and melons. Unlike many moths, they are fast daytime fliers that are sometimes mistaken for wasps.

Distribution

It occurs east of the Rocky Mountains from Canada to South America and India also.

Identification

Adults: The adults have stout bodies, about 3.8cm in length with a wingspan from 3.2 to 3.8cm. The adult body is reddish with black bands encircling the abdomen. The front wings are a metallic green. The hind wings are clear with dark veins and fringed with reddish brown hairs. Unlike many moths, it is a daytime flyer and can sometimes be seen hovering in front of flowers feeding on nectar or around cucurbit plants and vines.

The larvae are white to cream-colored caterpillars with brown heads which grow to about one inch in length and eggs are small, disk-shaped, reddish-brown in colour.

Biology

Female squash vine borer lay eggs on stems or leaf stalks at the base of plants or vines. Eggs take 9–11 days to hatch (Canhilal *et al.*, 2006), and larvae burrow into stems within hours of hatching (Welty, 2009) leaving small, almost invisible entrance holes and yellowish frass. Larvae are wrinkled, whitish worms with a brown head capsule. Larvae feed for approximately 25–27 days (Canhilal *et al.*, 2006). After feeding for about a month the borers exit from the stem and burrow about 1-6 inches deep into the soil for pupation. Pupation takes place in earthen cocoon in the soil. Overwintering takes place in the cocoon in pupal form during spring. One to two overlapping generations completed per annum.

Mode of feeding and symptoms of damage

The larvae cause damage to the plants by cutting the water and nutrient conducting lines between root and shoot. As a result, the plants start to wilt or leaves begin to turn yellow and eventually brown around the leaf margins. Diseases such as bacterial wilt or Fusarium wilt can cause similar symptoms. The primary feature for distinguishing squash vine borer from these two diseases is the presence of frass (insect excrement), which resembles wet sawdust, accumulating at the entrance of the larval tunnel. If these symptoms exist, split the stems apart with a sharp knife to look for the larvae. If several larvae found to infest a single plant, the plant may collapse and die.

Management

a) Disrupting mature larvae or pupae in the plants or soil by tilling in crop debris soon after harvest is the primary cultural practice for preventing a buildup of squash vine borer. This can be particularly important for crops such as summer squash where growers often make succession plantings and abandoned plantings may be left standing because of busy summer harvest schedules. Making the effort to till under abandoned plantings will help prevent buildup of damaging populations. These techniques are only suitable for very small plantings. Examine stems during peak egg-laying and crush or remove eggs before hatching.

b) *Steinernema carpocapsae* or *S. feltiae* applied to the stem and soil provided control similar to a conventional insecticide.

c) The stage most susceptible to natural enemies is the egg stage, which is attacked by parasitic wasps. Hence encouragement of parasitoid *Apanteles* spp. activity is beneficial.

d) Spray any insecticides viz. methyl demeton 25 EC @ 500 ml/ ha. Spray timing is critical for effective chemical control of squash vine borer because the larvae begin to tunnel into the stem within hours of hatching from eggs (Welty, 2009). Insecticide residues must be present at egg hatch so that larvae come in contact or feed on residues as they enter the stem. One should take control measure against the newly hatching larvae (first instars) before they enter the plant. Once the larvae enter within the stem, little can be done.

e) Spray any following insecticides *viz.* malathion 50 EC @ 500 ml, dimethoate 30 EC 500 ml, methyl demeton 25 EC @ 500 ml/ ha (agritech.tnau.ac.in, 2015).

f) Growers can employ either of two main chemical control strategies: The first strategy is based on weekly application of insecticide regardless of whether the stem borer has been detected or not. Spraying starts immediately after the vines begin to run and is continued for four to six weeks. This strategy is relatively simple to implement but the weekly application of insecticides results in excessive chemical use that can harm non-target organisms. In the second strategy, the crop is sprayed only after the population has reached the economic threshold (Brust, 2010).

The economic threshold level for stem borer is two or more individuals found per 20 m of crop row or caught in pheromone traps per week. The borer population is monitored early in the season by scouting and pheromone trapping of adult moths. This second strategy is as effective as the first strategy described earlier and it requires less pesticide. However, it is relatively labor intensive hard to implement and is only economically feasible when dealing with more than 5 acres of crop (Brust, 2010).

2.1.10 Flea beetle, *Phyllotreta cruciferae* (Goeze) (Chrysomellidae: Coleoptera)

Hosts of commercial importance

The pest is highly polyphagous in nature. It is a major pest of cruciferous vegetables and occasionally found to infest cucurbitaceous vegetables. Interestingly, the beetle prefer plant families that producc allyl isothiocyanate which is a known aggregation pheromone for the beetle.

Distribution

It has been reported from Europe, Egypt, Middle East, Asia, Russia and North America. In India, the species is commonly found on cole crops.

Identification

Adult beetles are 1.5-2.0mm. in length, elongate oval in shape and metallic bluish green in colour, antenna is black. Prothorax is wider than its length and scutellum is small. Elytron is closely covered with punctures. The beetles have enlarged hind femora which they use to jump quickly when disturbed. Their name, 'flea beetle' arose from this behavior. Eggs are yellow, oval, about 0.38-0.46mm. in length, 0.18-0.25mm. in wide deposited singly or in groups of 3-4 adjacent to the host plants' roots. Larvae are small, whitish, slender, cylindrical worms. They have tiny legs and a brown head and anal plate. Pupa are black in colour and free body appendages.

Biology

A single female can lay 50-80 eggs in her life time. The sexually mature female lay eggs near the host plants on soil. Incubation, grub, prepupal and pupal stages last for 5-10, 9-15, 2-4 and 8-14 days respectively depending upon environmental condition. The pest has been found to complete 7-8 overlapping generations per annum (Varma, 1961).

Mode of feeding and symptoms of damage

Immediately after hatching the grubs feed on roots. However, the damage to roots is not so serious on cucurbitaceous vegetables in India and no management practice is required to be adopted. The adult beetles feed on foliage and make typical 'shot hole' symptoms particularly during seedling stage of the crops.

Seasonal incidence

In India, the peak period of infestation of the pest is during March-April after which it migrates to root crops like raddish, turnip etc.

Management

a) It is a minor pest of cucurbitaceous vegetables and hence it usually cause no or meager damage to the crops.

b) Close vigil to be maintained in hot and dry weather, when flea population can rapidly increase.

c) Economic threshold level for foliar application of insecticides is 25% of the surface area, first two leaves has been injured and beetles are present. In that cases chlorpyriphos may be applied. If leaf damage is less than 25% and the crop is actively growing, the crop can generally recover.

d) However, if its attack is noticed during later stage of the crop, the crop may be sprayed with 0.2% carbaryl WP or dusted with carbaryl (Butani *et al.*, 1977).

e) Chemical control is recommended when flea beetle populations exceed threshold levels particularly early in the season. Commercial fields should be scouted for adults with an insect sweep net because flea beetles can move into a field quickly. Newly planted fields should be scouted for insects or damage every one to two days while plants are small and unable to withstand much damage. Soil-applied insecticides at planting will provide season-long control. Foliar insecticides provide quick control of large populations of adult flea beetles. When selecting foliar insecticides, choose products that will not disrupt natural enemies of other pests such as lepidoptera on cole crops.

2.1.11 Aphids, *Aphis gossypii* Glover. and *Myzus persicae* (Sulz.) (Aphididae: Hemiptera)

Aphis gossypii Glover. is known as melon aphid and *Myzus persicae* (Sulz.) is known as green peach aphid.

Commercial hosts

At least 60 host plants of *A. gossypi* are known from Florida and perhaps 700 host plants world-wide. It is a notorious pest of melons and a number of cucurbitaceous vegetables and is also found to infest various malvaceous and solanaceous plants (Srivastava and Butani, 2009). Other vegetable crops seriously affected are asparagus, pepper, eggplant, and okra. Some other important crops injured regularly are citrus, cotton, and hibiscus. *Myzus persicae,* being a highly polyphagous pest, in addition to infest cucurbits also attacks peach and other temperate fruits.

Distribution

It has a very wide cosmopolitan distribution except in extremely cold region of Asia and Canada.

Identification

Both nymphs and adults are soft bodied, delicate and pear shaped insect. The adults of *A. gossypii* are wingless (apterous), parthenogenetic females are 1 to 2 mm in length. The body is quite variable in color, light green mottled with dark green is most common, but also occurring are whitish, yellow, pale green and dark green forms. The adult aphid varies in color from light to dark green and has dark siphunculi (Slosser *et al.*, 1989). At high temperatures and in crowded environments, adult aphids are pale yellow in color (Blackman and Eastop, 1985). The legs are pale with the tips of the tibiae and tarsi black. The cornicles also are black. Small yellow forms apparently are produced in response to crowding or plant stress. Winged (alate) parthenogenetic females measure about 1.1 to 1.7 mm in length. The head and thorax are black and the abdomen yellowish green except for the tip of the abdomen, which is darker. The wing veins are brown. The egg-laying (oviparous) female is dark purplish green; the male is also similar.

Myzus persicae

Nymphs initially are greenish, but soon turn yellowish, greatly resembling viviparous (parthenogenetic, nymph-producing) adults. Females gave birth to offspring 6 to 17 days after birth, with an average age of 10.8 days at first birth. The length of reproduction varied considerably, but averaged 14.8 days. The average length of life was about 23 days, but this was under caged conditions where predators were excluded. The daily rate of reproduction averaged 1.6 nymphs per female. MacGillivray and Anderson (1958) reported five instars with a mean development time of 2.4, 1.8, 2.0, 2.1 and 0.7 days respectively. Further, they reported a mean reproductive period of 20 days, mean total longevity of 41 days and mean fecundity of 75 offspring.

Winged (alate) aphids have a black head and thorax and a yellowish green abdomen with a large dark patch dorsally. They measure 1.8 to 2.1 mm in length. Winged green peach aphids seemingly attempt to colonize nearly all plants available. They often deposit a few young and then again take flight. This highly dispersive nature contributes significantly to their effectiveness as vectors of plant viruses.

The cornicles are moderately long, unevenly swollen along their length and match the body in color. There exists both alate and apterous forms in a single population. However, the apterous forms are more common and abundant.

Biology

The life cycle of the pests is very much interesting and complex. Mode of reproduction is usually viviparous and parthenogenetic and sometimes oviparous. The mature female directly gives birth to offspring in place of laying eggs and that too without fertilization. A single apterous female produces 8-22 nymphs per day. These mature within 3-4 days and total life cycle occupies about 7-9 days. The alate forms generally appear after several generations, especially when there is overcrowding, deterioration of plant condition and/or change in the environmental condition particularly the temperature. Parthenogenetic reproduction allows the overlapping generations and a higher rate of population increase (Dixon, 1987). The nymphs that give rise to winged females (alatae) may be pinkish. The dispersants typically produce about 20 offspring, which are always wingless. This cycle is repeated throughout the period of favorable weather.

Mode of feeding and symptoms of damage

The aphids are gregarious feeder and large colonies of nymphs and adults are found to infest tender twigs, shoot as well as lower leaf surface and suck cell sap therefrom. Due to sucking sap the plant turn yellow, leaves become curled, wrinkled and deformed in shape and ultimately dry. Vigour and vitality of the plant get affected badly. Fruit size, shape and quality are also reduced in case of a bearing plant. Heavy infestation of aphids also results in secretion of copious quantities of honey dew on the infested parts on which sooty mould fungus (*Capnodium* spp.) develops. The amount of honeydew produced by aphids in the 1st instar is significantly higher than that of other instars (Henneberry *et al.*, 2000). Development of sooty mould make the plant surface covered with a black thick layer that ultimately hinder photosynthetic activity of the vines and retard growth resulting stunting of the infested plant. The infested fruits covered with black sooty mould fetch poor market price.

Besides these direct damages caused by desapping the vines, the aphid also transmits viral disease and the loss due to this disease is irrepairable. *A. gossypii* alone found to be a vector of 76 virus diseases in a wide range of plants (Chan *et al.*, 1991). The main disease which is transmitted by the aphid on cucurbits is Cucumber Mosaic Virus (CMV). Other important diseases of different crops transmitted by this aphid are Bean Common Mosaic (BCM), Bean Yellow Mosaic (BYM), Lettuce Mosaic (LM), Beet Mosaic (BM), Cabbage Black Ring Spot (CBRS) and Citrus Tristeza.

Management

The wide host range of melon aphid makes crop rotation a difficult tactic to implement successfully. Also, crops grown down-wind from infested fields are especially susceptible because aphids are weak fliers and tend to be blown about.

a. Cultural practices

i) Infested crops should be destroyed immediately after harvest to prevent excessive dispersal and it may be possible to destroy overwintering hosts if they are weeds.

ii) If continuous cropping is implicated in retention of aphid populations then a crop-free period is needed. Row covers can be used to inhibit development of aphid populations.

iii) Time of planting may influence potential aphid population increase.

iv) Yellow sticky traps to be installed.

b. Insecticides

i) If insecticides are used to suppress melon aphid, care should be taken to obtain thorough cover of foliage.

ii) Leaf distortions caused by aphid feeding provide excellent shelter for the insects, so systemic insecticides are useful.

iii) Excessive and unnecessary use of insecticides should be avoided.

iv) Early in the season, aphid infestations are often spotty, and if such plants or areas are treated in a timely manner, great damage can be prevented later in the season.

v) Use of insecticides for other more damaging insects sometimes leads to outbreaks of melon aphid.

vi) Melon aphids will transmit viruses to crops that they do not colonize. Insecticides have little effect on virus transmission by non-colonizing, transient aphids, though insecticides can prevent secondary transmission within crops where colonization occurs.

vii) Foliar spraying with confidor 17.8 SL (Imidacloprid) and Actara 25WG (Thiamethoxam) was found to be effective as reported by Khan *et al.* (2011).

viii) Several systemic insecticides, including dimethoate 30 EC @ 1.7 ml/ l adequately control aphids and can be used if needed (NCIPM, 2010).

Management of disease transmission

a) It is difficult to disrupt transmission of nonpersistent viruses with insecticides, so total dependence on insecticides is not advised to check viral disease transmission.

b) Row covers and reflective mulches or coarse net covers are helpful in delaying or reducing disease transmission, but this is expensive to implement on a large scale.

c) Both aluminum and plastic mulch are reported to be useful for suppression of watermelon mosaic virus. Elimination of alternate hosts of both aphids and the virus diseases is often key to disease management; both weeds and crop plants can harbor the disease and vectors.

d) Transmission of nonpersistent viruses such as cucumber mosaic virus can sometimes be reduced by coating the foliage with vegetable or mineral oil. Oil is postulated to inhibit virus acquisition and transmission by preventing virus attachment to the aphid's mouthparts, or to reduce probing behavior. Oil seems to be the most effective when the amount of disease in an area that is available to be transmitted to a crop is at a low level. When disease inoculum or aphid densities are at high levels, oils may be inadequate protection. However, some plants may be damaged by oil applications, especially during hot weather.

2.1.12 Stink bugs, *Apspongopus brunneus* Thunberg. *A. janus* (Fab.), *A. observus* Fab. (Pentatomidae: Hemiptera)

Hosts of commercial importance:

The pests have often found to infest cucurbitaceous vegetables, specially pumpkin and various other gourds.

Distribution

The bugs are distributed all over Indian sub-continent. *Apspongopus brunneus* is usually found in south India, *A. observus* found mainly in North-East India and *A. janus* is distributed widely all over India.

Identification

The adults are flat, medium sized insects.

Apspongopus janus: The adults of the insect is about 30mm. in length, pronotum and base of the elytra is bright red while the head and wing membrane are black.

Apspongopus brunneus: Pale brown in colour and slightly smaller in size.

Biology

The sexually mature females deposit their eggs in long rows on tender leaves. Incubation period and larval periods are 9-10 and 24-28 days respectively depending upon prevailing environmental conditions.

Mode of feeding and symptoms of damage

Clusters of both nymphs and adults may be found clinging to the leaves and tender shoots of the plants. They suck cell sap and thereby devitalize plants and retard their growth and development. The bugs emit characteristic buggy smell and hence the common name stink bug.

Management

a) During early stages of infestation collection and manual destruction of the leaves and twigs containing congregating bugs is effective to suppress the population build up.

b) Application of organic pesticide like neem oil, insecticidal soap, pyrethrin and rotenone etc may safely be recommended.

c) Cypermethrin sprays can sometimes be effective and are easily degraded in soil as well as on plants.

2.1.13 Leaf footed plant bug, *Leptoglossus australis* (Fab.) (Coreidae: Hemiptera)

Hosts of commercial importance

It is a polyphagous pest but the main hosts are cucurbitaceous crops. Among the cucurbits bitter gourd is the most preferred one (Fernendo, 1957; Jadhav *et al.*, 1979). In addition to that it has also been found feeding on cotton and orange in Malaysia (Pagden, 1928) citrus and tomato in Sri Lanka (Hutson, 1936), passion fruit in Kenya (Hill, 1975), banana, citrus, pomegranate and passion fruit (Jadhav *et al.*, 1979) and snake gourd (Visalakshi *et al.*, 1980) in India. The pest also found to infest several other crops viz. cocoa, coffee, groundnut, oil palm, sweet potato, yam etc. (Butani and Jotwani, 1984).

Distribution

The pest is known from the tropics and sub tropics throughout the world including Northern Africa (Packauskas and Schaefer, 2001). It has also been recorded from Western, Central, Eastern Africa, South China, Phillipines, Indonesia, New Guinea, Pacific Islands and North Australia (Srivastava and Butani, 2009).

Identification

Adults: The adult insects are about 20mm. long, black coloured bugs. Antenna 4 segmented, alternately with black and yellow colour. Mesonotum is laterally produced into a spine on either side. Minute yellow spot is present in the centre of each forewings and on hind tibia.

Nymphs: Eggs are cylindrical in shape and light brown in colour. Full grown nymphs are 12 to 16mm. long and blackish in colour. Head is characterized by two black spines dorsally in between the compound eyes. Thoracic plate carries a spine laterally on either side and abdominal segments are also protruded laterally into spines. Wing pads appear after third instar when hind tibia also flattened out as leaf like structure and due to which the pest is named so.

Biology

The eggs hatched 6-7 days after oviposition. The newly hatched nymphs passes through five nymphal stages to become adult. The nymphal stage lasted about 52 days. The adult life-span is about 14 days for males and 10 days for females.

Mode of feeding and symptoms of damage

Both nymphs and adults of the pest cause damage to the plant by sucking the plant sap from terminal shoots and young fruits. The bugs congregate on terminal shoots and feed therefrom. Affected shoots withered beyond the point of attack while the affected fruits gradually turned yellow, shriveled and fall prematurely. Sometimes the affected fruits develop dark spots at point where the punctures are made.

Though the pest is of quite common occurrence to the damage is seldom severe i.e. crosses the threshold level of infestation. So, management practices for the pest are seldom required.

Seasonality of occurrence

In Japan overwintered adults of *L. australis* bred from May to June. In this period of time, the population density increased due to the lower occurrence of natural enemies such as an egg parasite, *Gryon pennsylvanicum* and ant species (Yasuda, 1989).

Management

a) To check the pest population 0.2% carbaryl may be sprayed.

b) The most effective insecticides against leaf footed bug are broad-spectrum, pyrethroid-based insecticides. However, these products are quite toxic to bees and beneficial insects. Insecticidal soap or botanicals, such as neem oil or pyrethrin, may provide some control of young nymphs only. If insecticides are used close to harvest, make sure to observe the days-to-harvest period indicated on the insecticide label; and wash the fruit before eating (UC IPM Pest Management Guidelines: Cucurbits, 2016).

2.1.14 Stem boring beetles, *Apomecyna* spp. (Lamiidae: Coleoptera)

Hosts of commercial importance

A number of longicorn beetles viz., *Apomecyna saltator* Fab., *Apomecyna alboguttata* Thomson, *Apomecyna histrio* Fab. and *Apomecyna perotetti* Fab. are found to infest cucurbitaceous vegetables. The better known large species are mainly the pest of forest trees as their grubs can bore in hard woody tissues.

Distribution

The cucurbit longicorns are widely distributed species in the Afrotropical, Oriental, Australasian regions (May, 1946), Hawaii, and India (Samalo and Parida, 1983; Samalo, 1985). Among the aforementioned species *A. saltator* is found all over India whereas *A. perotetti* is more common in southern tract of India. On the other hand *A. histrio* has been recorded only from Bihar state.

Identification

Adults

Apomecyna saltator: Adults are about 12-18mm. long, brown coloured beetles, densely pubescent dorsally and sparsely on the underside of the body (ventrally). The beetle is slate-gray in colour, 3.45 and 3.04 mm in breadth across thorax and abdomen respectively with long antennae about two-third the length of the body and for this reason this is also known as longicorn beetle. Head is bent downwardly, thorax and elytra are marked with a number small to larger white spots. These spots run into each other to appear as large white irregular patches on each elytron, one is little below the base and the other between the apex and the middle region.

A. alboguttata: These are to some extent smaller in size, about 8-10mm. in length, dark brown in colour with round white spots on thorax and elytra.

A. histrio: It has a pair of white spots similar to those of *A. saltator* but differ in the design of elytron. At the base of each elytra there is an oblique row of four white spots between the middle and apex. Adults are able to make cocophony (sharp squeaking noise) by rubbing the inner surface of hind margin of pronotum on a small oval shaped, smooth and highly polished raised surface on mid-dorsal region of mesothorax (Srivastava and Butani, 2009).

Grubs: Grubs are brownish in colour having flattened head and thorax. Abdomen is soft and distinctly segmented and when full grown the grubs measure about 18-20mm. and breadth across thorax and abdomen are 2.69 and 2.66 mm. respectively.

Pupa: Pupae are pale yellowish in colour and 10-13mm. in length and breadth across thorax and abdomen are 3.28 and 2.71 mm. respectively. Type of pupa is exarate.

Eggs: Eggs are elongated oval in shape, about 1.5mm. in length and creamy white in colour with reddish tinge.

Biology

The biology of this beetle has not been studied well and the duration of each life stage is not known (Khan, 2012) in Bangladesh condition. However, Lefroy (1910) found incubation, grub and pupal period of the beetle as 5-6, 22-23 and 6-8 days respectively. Eggs are laid by the sexually mature female singly in the epidermis of the stems of cucurbit vines. On hatching the grubs bore into the long trailing stems usually at or near node and tunnel inside along the central pith. The larva pupate within a fibrous cocoon constructing within these tunnels. The adult beetles emerge by biting their way out of the stem. Total life cycle comprises 35-46 days and adult longevity is about 33-39 days. Three to four generations per annum was stated by Lefroy (1910). However, May (1946) reported that the cucurbit longicorn may have two or three generations per year in melons in Queensland, Australia. The pest overwinters as grub inside the tunnel of infested vine of crop.

Mode of feeding and symptoms of damage

Immediately after hatching the grubs bore into the long trailing stems with their biting mouthparts usually at or near node and tunnel inside along the central pith eating away the surrounding tissues. Grubs bore into the main vines of plants and produce a swelling of the stem. The feeding tunnel is usually directed towards nodes and is filled with glutinous waste material (May, 1946). The larvae enter the stem and infested plants usually show no conspicuous symptoms. Under very severe infestations, young plants may die, but older plants often live to produce reduced yields. The full grown larva pupate within the tunnel in a fibrous cocoon. The adult beetles emerge by biting their way out of the stem and feed by gnawing the leaf petioles and softer parts of the stems. Occasionally, the leaves are also eaten making irregular hole on leaf lamina. Though the beetle occurs quite often, but the damage caused is seldom severe in India.

Seasonal incidence

After passing overwintering emergence of adults from dried vines takes place usually during May. Mating takes place immediately after emergence. During mating the females move about freely and keep on feeding, carrying the copulating males on their back. They become active and found to infest crops during summer and monsoon months. In extreme cold months of the year go for hibernation in grub stage.

Management

a) In Indian condition the pest is a minor one for cucurbitaceous crops. No control measures are usually adopted against the beetle.

b) Cutting and destroying of affected vines along with the pest is a viable option as suggested by Butani and Jotwani (1980) which will prevent population build up and spread of the pest.

2.1.15 Blister beetle, *Mylabris pastulata* Thunberg. (Meloidae: Coleoptera)

Hosts of commercial importance

Important hosts of the pest are various cucurbits including melons. It is also found feeding on cotton, ground nut, millets, okra, rose etc. The adult beetles are phytophagous and it has been observed feeding on plant materials, particularly of alfalfa, peanuts, soybeans and many other plant species (Ward, 1985). The Meloids have also been detected feeding on a wide range of host plants within families, particularly Asteraceae, Leguminosae, Compositae, Umbeliferae, Solanaceae Fabaceae, Malvaceae, Convolvulaceae and Solanaceae (Selander, 1986; Arnett *et al.*, 2002; Bologna and Pinto, 2002; Lebesa *et al.*, 2011). Although the presence of blister beetles in different crops is usually not considered to be a serious pest (Zhu *et al.*, 2005). However, Hall (1984) and Nikbakhtzadeh (2004) reported that infestations of crops by the beetle only cause considerable damage because of its gregarious nature of feeding.

The Chinese blister beetle *Mylabris phalerata* (Pallas) was recorded as a serious pest of pigeonpea in the lower hills of Uttar Pradesh, India (Dutta and Singh, 1989). The blister beetles *Zonabris pustulata* Thunb. and *M. pustulata* were mentioned as pests on Cashew in Andhra Pradesh (Ayyanna and Ramadevi, 1987; Sreedevi *et al.*, 2009).

Distribution

It is a polyphagous pest widely distributed in Africa and South-East Asia (Hill, 1975).

Identification

Adults: Polyphagous beetles feeding on crops with yellow or pink flowers. Adults are medium sized beetles with black head, thorax and abdomen and are about 22-36mm. or more than that in length having three black and three yellowish orange bands running vertically and alternately on elytra. The insects when handled or disturbed exude an acrid yellow fluid which contains cantharidine. Cantharidine is an irritant to skin and causes blisters on human skin and hence it is known as blister beetle.

Cantharidin (as a haemolymph exudation) serves as a feeding deterrent to most predators, thereby protecting blister beetles and their eggs from predation. However, some insects are attracted to cantharidin, and this compound is involved in the chemical communication among blister beetles (Young, 1984; Klahn, 1987). Because of the poisonous nature of cantharidin, these beetles periodically and inadvertently eaten (with feed such as hay) by the domestic livestock and horses causing severe illness or death.

Biology

Eggs are laid by the adult female in soils of cultivated fields. On hatching the grubs feed on egg pods of various grasshoppers and locusts found in the soil. Thus this stage of the pest can safely be considered as highly beneficial from pest management point of view (Singh, 1970). Full grown larvae are coarctate and form pseudopupae that become pupae later on. Pseudopupae are devoid of functional appendages and hibernation takes place in this stage in soil.

Mode of feeding and symptoms of damage

Main damage to crops is caused by the adults through their cutting and chewing type of mouthparts. Adult beetles are polyphagous and found attracted to yellow and pink flowers. They emerge out of the soil around August and are active till early December (Srivastava and Butani, 2009). However, peak activity is usually observed from August -October in South India. They feed on pollens, petals of flowers and flower buds. The resultant effect is the poor fruit setting. In addition to cucurbits, considerable yield losses caused by blister beetle, *Mylabris* spp. in pigeonpea also.

Management

It is difficult to control this beetle because of their rapid mobility. However, hand collection and prompt destruction of the beetle can keep the population under check. This may be done during early morning hours when the beetles remain comparatively less active.

a) Manipulation of cultural practices to avoid pests or to make the environment less favorable for them. There are several commonly used methods such as crop rotation, sanitation, polyculture, strip cropping and trap cropping etc.

b) Regular intensive weeding and early crop establishment helps in reducing this pest population build up.

c) Reducing weedy host plants and harvesting prior to bloom are sound management tactics.

d) In the present century, bioefficacy studies for the insecticides Beta-cyfluthrin, deltamethrin were carried out on the sponge gourd against blister beetle, *M. pustulata* in India and found effective.

e) Little is known concerning the natural enemies of blister beetles. Undoubtedly, starvation of first instars is a very important factor during most seasons, and cannibalism is prevalent among larvae. Ant-like flower beetles (Coleoptera: Anthicidae), false ant-like flower beetles (Coleoptera: Pedilidae), and some plant bugs (Hemiptera: Miridae) have been implicated as mortality agents of blister beetles (Selander, 1986).

f) *Beauveria bassiana* Bals. Vuillemin is the most popular among the registered mycoinsecticides. One of the principal reasons for its popularity is its very wide host range of about 750 insect species (Khan *et al.*, 2005).

g) Most insecticides are not very effective against these beetles, but synthetic pyrethroids such as cypermethrin 10 EC @ 1.0 ml/l or lamda cyhalothrin 5 EC @ 1.0 ml/l work reasonably well (NCIPM, 2010).

2.1.16 Melon thrips, *Thrips palmi* Karny (Thripidae: Thysanoptera)

Hosts of commercial importance

Tobacco, cotton, cucurbits, crucifers, egg plant, beans, soybean, peanut, avocado and mango etc.

Distribution

Asia, Pacific, Carribean, Central and South America, Florida (Capinera, 2001).

Identification

Adults are pale yellow in colour, females are about 1.16mm in length and the males 0.91mm. Eggs are laid on young leaves or flower buds. Incubation period is about 4 days. There are two nymphal stages of the pest and damage is caused by both nymphs and adults actively feeding on the tender parts of the plants particularly on leaves and buds. After second instar they undergo prepupal and pupal instar in soil.Total life cycle is completed in about 11-14 days in the tropics.

Mode of feeding and symptoms of damage

Feeding by both nymphs and adults by rasping and sucking mouthparts result in dropping of flower buds, stunting of terminal shoots and curled and scarred fruits. The most serious but indirect damage caused by the pest is transmission of tospoviruses. Peanut necrosis virus on peanut, potato, soybean and sunflower, Capsicum chlorosis virus on peppers etc.

Management

Predatory bugs *Orius mexidentex* and *Carayonocoris indicus* have been reported to feed on *T. palmi* in India.

In addition to the aforementioned insect pests, some others are also reported to infest cucurbitaceous vegetables in different cucurbit growing regions of the world. However, they are not responsible for heavy crop loss and even no control measures are required to adopt for their management. The injury that they inflict is minor and sometimes the damage is recoverable especially in the cases where the plant is in active growing stage. In some instances they act as sporadic pests and do not cause detectable injury to the plant. Some of them are Water-melon weevil, *Acythopius citrulli* Marshal (Curculionidae: Coleoptera), Gray weevil, *Mylloceros blandus* Faust (Curculionidae), Leafhoppers (Jassids), *Eutettix phycitis* Dist. (Cicadellidae: Hemiptera), Mirid bugs, *Creontiades palidifer* Walk. (Miridae: Hemiptera), *Cyrtopeltis tennis* (Reuter). A small dark weevil, *Bans trichosanthis* Subra. has also been reported feeding on tender shoots of snake gourd in South India.

2.2 Mite Pests

The phytophagous mites belong to four important families *viz.*, Tetranychidae (spider mites), Tenuipalpidae (false spider mites), Eriophyidae (gall mites) and Tarsonemidae (broad/yellow mites). Out of 37 mite species known to feed upon vegetable crops, six species are serious pests mostly of vegetables like brinjal, okra, cucurbit, chilli, potato etc. in major parts of the country (Gupta, 1991). The red spider mite, *Tetranychus neocaledonicus* Andre. is one of the most destructive pests of cucurbits especially in warmer climate. In vegetables spider mite damage alone causes 10 to 15 per cent loss in yield (Anon., 1991b). In recent years, the problem of phytophagous mites has gained momentum due to development of resistance and resurgence (Vande Vrie, 1985).

2.2.1 Red Spider Mite, *Tetranychus neocaledonicus* Andre. (Tetranychidae: Acarina)

Hosts of commercial importance

In addition to cucurbits the other hosts of commercial importance include beans, egg plant, cole crops, okra, onion, peas, potato, sweet potato, tamarind, tomato and some other ornamental plants.

Identification

Spider mites are very tiny, the adult females are only 0.3 - 0.5 mm long and the males are even smaller. They are often red, brown, green or yellow in color and they can be seen clearly with the aid of a lens. Abdominal segmentation is distinct and there are twelve pairs of hair on the dorsum. There also remain two black spots on the dorsum that increase in size with age and finally cover entire dorsum.

Eggs are spherical, minute (about 0.1 mm in diameter) and transluscent. Freshly hatched larva are almost spherical in shape, 0.1 to 0.2 mm in diameter and light amber coloured. Later these become elongated in shape and greenish colour. They have three pairs of legs and two small dark specks dorsolaterally one on either side. The immature stages resemble the adults except in size. Protonymphs are deep green in colour and slightly larger than the larvae (0.2 to 0.3 mm in diameter) with longer bristle on dorsum and four pairs of legs. Males are elongate in shape and the females are ovate. Deutonymphs are slightly bigger than the protonymphs and are only found in females. In case of deutonymphs the genitalia is well visible.

Biology

Eggs are laid by the adult females on the webbings of ventral surface of leaves. Pre-oviposition and oviposition period varies from 1-2 and 8-12 days respectively during March to September. Oviposition period may go up to 20 days during cooler months i.e. during October –December. Fertilized female lay more eggs (61-93) than the unfertilized one (39-59).

Duration of different developmental stages varies depending upon prevailing environmental condition. Incubation period ranged from 2-30 days. Larval and nymphal period takes 1-6 and 2-9 days during May to October in case of males and females respectively. They require more time during cooler months i.e. during November to February. Total life cycle is completed in 4-8 and 6-10 days during April to October in case of males and females respectively. In the development of its life cycle parthenogenesis is common. However in that case the progeny will contain only the male individuals. Under controlled conditions up to 32 generations are observed per annum.

Mode of feeding and symptoms of damage

Spider mites have needle-like mouthparts and feed by piercing the leaves of host plants and sucking out the fluids from plant cells. Both mature and immature stages feed by sucking the plant sap from tender parts of plant. Colonies of mites can be seen on ventral surface of tender leaves protected under silken webs. Desapping make the infested plant weak. The plant surface is covered with thick webs on which soil particles are collected during dry and windy weather sometimes results in leaf dropping that ultimately affect growth, flowering and fruit formation of the plant.

Feeding causes yellow spots on the leaves and in heavy infestations, foliage shows a yellowing or bronzing appearance and may suffer from premature leaf drop. In situations where there is severe damage this may lead to plant death. Mites prefer the young leaves, however in heavy infestations, the older leaves are also affected and sometimes webbing may be seen all over the plant.

Peak population is usually observed during post monsoon period and their activity declines with the fall in temperature. It is interesting to note that all developmental stages of the mite is destroyed during rainy season except the eggs and hence occurs rapid multiplication of spider mite after rainy season that continued up to following winter.

Management

a) Timely inspection of susceptible plants is important in preventing serious damage.

b) Regular monitoring of plants that have a history of mite problems. The best method of monitoring mites is to walk through the crop randomly checking the underside of new and medium aged leaves on a weekly basis.

c) Looking out for yellow or distorted leaves as these may be symptoms of mite infestation. The regular use of miticides may kill predatory mites or create problems with pesticide resistance in the plant feeding mites.

d) Soft chemical sprays such as petroleum oil and potassium soap (2ml/lt of water) are effective in controlling certain species of mites in crops.

e) Biological control of spider mites has been successful in some countries. In Australia, for example, the predatory mites, *Phytoseiulus persimilis* Athias-Henriot and *Typhlodromus occidentalis* Nesbitt have been produced commercially for the purpose of controlling some important spider mites, such as *Tetranychus* spp. on vegetables, ornamental plants and fruit trees.

2.2.2 Oriental red mite, *Eutetranychus orientalis* (Klein) (Tetranychidae: Acarina)

Distribution

This species has a wide distribution in the whole World but it is not clear which is the country of origin. In India it is distributed in Assam, Delhi, Haryana, Karnataka, Kerala, Madhya Pradesh, Meghalaya, Punjab, Rajasthan, Tamil Nadu and Uttar Pradesh.

Host range

Citrus spp. are the main hosts of economic importance. The mite has been recorded on a wide range of other crops including almonds (*Prunus dulcis*), bananas (*Musa paradisiaca*), cassava (*Manihot esculenta*), cotton (*Gossypium*), figs (*Ficus carica*), guavas (*Psidium guajava*), mulberries (*Morus*), olives (*Olea europaea*), pawpaws (*Carica papaya*), peaches (*Prunus persica*), pears (*Pyrus*), *Plumeria*, quinces (*Cydonia oblonga*), *Ricinus communis*, sunflowers (*Helianthus annuus*), sweet potatoes (*Ipomoea batatas*), watermelons (*Citrullus lanatus*) and over 50 other plant species.

Identification

Eggs are oval or circular and flattened, coming to a point dorsally but lacking the long dorsal stalk of other spider mites. Newly laid eggs are bright and hyaline but later take on a yellow, parchment-like colour (Smith-Meyer, 1981). Average size of larva is 190 x 120 µm. The protonymph is pale-brown to light-green, legs shorter than the body, average size 240 x 140 µm. The deutonymph is pale-brown to light-green, average size 300 x 220 µm. Adult female is broad oval, flattened, varying in colour from pale-brown through brownish-green to dark-green with darker spots within the body; legs about as long as the body and yellow-brown. Average size is 410 x 280 µm. Male much smaller than female and elongate triangular in shape with long legs (leg about 1.5 x body length).

Biology

In 1988-89 in Jordan on lemons, mites were most abundant during late November and December, a season of lower temperatures and increased rainfall. The mites passed late January through February as mature, quiescent fertilized orange-red females. There were 8-10 generations per year (Tanigoshi *et al.*, 1990). In India, there are usually two population peaks, early May and mid-September, rains during July and August causing a significant reduction in numbers.

Damage symptoms

Mites commence feeding on the upper side of the leaf along the midrib and then spread to the lateral veins, causing the leaves to become chlorotic. Pale-yellow streaks develop along the midrib and veins. Little webbing is produced. In heavier infestations, the mites feed and oviposit over the whole upper surface of the leaf. Very heavy infestations on citrus cause leaf fall and die-back of branches which may result in defoliated trees. Lower populations in dry areas can produce the same effect.

Management

a) Acaricides are commonly used against *E. orientalis* and other spider mites on citrus, flubenzimine and omethoate were effective (Sharaf, 1989).

b) In India, dicofol (0.025%) and sulfur (0.15%) were found to be the most effective against *E. orientalis*.

c) Spiromesifen may also be recommended for management of this mite pest.

2.3 Nematode Pests

2.3.1 Root-knot nematodes, *Meloidogyne incognita*

The root-knot nematodes cause root galls from the initial stages of crop growth. The larvae feed on the roots, which show typical galls, and later the entire root system shows heavy galling. The foliage becomes light yellowish; the plants become stunted and results in flower and fruit drop. In spite of irrigation the plants appear sick and drooping during daytime.

Hosts of commercial importance

Very broad host range. More than 700 hosts have so far been detected. These include most cultivated crops and ornamentals.

Distribution

The nematode is more important in warm temperate, tropical and subtropical regions of the world (Brodie *et al.*, 1993). The cosmopolitan distribution of some species is the result of the movement of rooted plants in commerce and at the local level through movement of water, soil and equipment and rooted seedlings of crop plants and ornamentals.

Biology

Sedentary endoparasite, parthenogenic in nature. Second-stage infective juveniles hatch from the eggs. These invade roots in the region of elongation near the root cap. They migrate between and through cells and position themselves with the head in the vascular tissues. Cell damage occurs as a result of the migration and if several juveniles enter the root tip cell division stops and there is no root elongation.

As feeding continues several cells near the head begin to enlarge and become multinucleate. These are called giant cells and there are usually 3 to 6 associated with each nematode. The formation of giant cells and galls is the result of cell enlargement and of increased numbers. These changes are induced by substances (salivary secretions) introduced into cells and surrounding tissues during feeding of the nematode. During this process the xylem vessels become disrupted and the roots cannot function normally with respect to water and nutrients.

During the process of gall formation the nematodes undergo the second, third and fourth molts to reach the adult stages. Mature females are saccate (pear-shaped) and lay eggs into a gelatinous matrix. This matrix may protrude from the surface of small roots or may be entirely within the gall. Eggs hatch in about 7 days. The entire life cycle is completed in 20-25 days at 70^0 F. Males are vermiform, not required in reproduction.

Symptoms of infestation

Reduced root systems and galling due to infestation. There is poor top growth and the foliage is frequently chlorotic (yellow) because essential elements are not taken in and transported by the impaired root system. Growth of the plant becomes stunted. Severe infections cause wilting of the foliage and the plants require more frequent irrigations. Muskmelon, cucumber, pumpkin etc. cucurbits are affected severely.

Management

a) Seed treatment with bio-pesticide like *Pseudomonas fluorescens* (@ 10g/kg seed).

b) Long crop rotation with non host crop.

c) Soil application with Nemagon.

d) Application of FYM (two tonnes) enriched with *Pochonia chlamydosporia* and *Paecilomyces lilacinus* per acre before sowing along with 100-200 kg of neem or pongamia cake is effective.

e) Soil solarisation and FYM (@4 t/ha) application reduces incidence of *M. javanica* on cucumber roots (Nasr-Esfahana and Ahmadi, 1997).

f) Green manuring with *Tagetes* or *Xanthium* leaf powder followed by application of *Verbesina* and *Artemesia* reduces population of *M. incognita* in cultivation of *C. pepo* (Sharma *et al.,* 1985).

g) Resistance to *M. incognita* has been observed in wild species *Cucumis dipsaceus* and *C. anguria* and *C. sativus* cultivars Fem Cap, Rozental Tsepellin Superator etc. (Udalova and Prikhod'ko, 1985).

In addition to that *M. hapla and M. javanica* have also been found to infest cucurbits in varying intensity (Srinivasan and Pal, 1998).

2.4 Parasitic Plant

2.4.1 Orobanche, *Orobanche ramosa* and *O. aegyptiaca*

The plant which is parasitic in nature is a species of broomrape known by the common names hemp broomrape and branching broomrape. It is a pest in agricultural fields, infesting crops like cucurbits, tobacco, potato and tomato etc. It is native to North Africa, but it is also known in many other places as an introduced species. From a thick root system the plant produces many slender and erect stems. The yellowish stems grow 10 to 60 centimeters tall and are provided with glandular hairs.

The broomrape is an annual and sometimes perennial plant and is parasitic on other plants, draining nutrients from their roots. It lacks leaves and chlorophyll. The inflorescence bears several flowers, each in a yellowish calyx of sepals and with a tubular white and blue to purple corolla.

Branched broomrape blooms from October to November. Flowers resemble small snapdragons; color varies from white to blue or violet. Twenty or more flowers cluster together to form a spike-shaped flower head. Upper flowers are stalkless and lower flowers are with short stalk. Stems and flower heads are covered with very short glandular hairs.

Management

a) Many management strategies have been tried against *Orobanche ramosa* and other broom rapes, but few of them have proved reliable and are only economical in high value agriculture.

b) The strength of branched broom rape lies in its ability to form a bank of seeds in the soil. A management or eradication program must aim at reducing this seed bank, while minimizing the production of new seeds and their dispersal to new sites.

c) Quarantine is therefore an essential element in management or eradication programme.

d) Integrated control using several techniques including those aimed at reducing seed recruitment are therefore advocated.

e) Combination of solarisation, herbicides and hand weeding with careful choice of cultivars and sowing times to manage *Orobanche*. However, none of these methods gave complete control when used separately.

f) The late post-emergence application of glyphosate was effective (Lolas, 1986) on *O. ramosa* in tobacco. Two treatments of 50 g a. i./ ha glyphosate reduced the number of *O. ramosa* and increased celery yield (Americanos, 1991).

2.5 Diseases of Cucurbitaceous Crops

Cucurbitaceous vegetables also suffer from the threat of many deadly diseases that considerably affect yield both quantitatively as well as qualitatively. Some of the important diseases are damping off, downy mildew, powdery mildew, gummosis, *Phytophthora* blight, anthracnose, *Cercospora* leaf spot, phoma blight, collar rot, charcoal rot, *Fusarium* wilt, white rot, root knot nematode, bacterial wilt, water melon bud necrosis and leaf distortion virus, mosaic, phyllody etc. These diseases are of national importance and cause significant economic losses in cucurbits. According to Pandey *et al.* (2002), diseases like gummosis, anthracnose, *Phytophthora* blight, *Cercospora* leaf spot, root knot nematode and water melon bud necrosis virus are becoming the most destructive diseases of cucurbits in todays agriculture. Name of different diseases along with their causal organisms are presented in Table 2, 3, 4 & 5.

Table 2. Diseases caused by the fungi and fungus-like organisms in field condition

S.No.	Name of the disease	Causal organism
1.	Seed rot and damping-off	*Pythium aphanidermatum* (Edson) Fitzp., *P. debaryanum* Hesse, *P. myriotylum* Drechsler, *P. butleri* Subram., *Rhizoctonia solani* Kuhn, *R. bataticola* (Tuabenh) Butler, *Phytophthora parasitica* Dastur, *Fusarium* spp., *Fusarium equiseti* (Corda) Sacc., *Acremonium* spp., *Thielaviopsis basicola* (Berk. & Br.) Ferraris and some other fungi
2.	Powdery mildew	*Erysiphe cichoracearum* DC., *Sphaerotheca fuliginea* (Schl.) Salmon
3.	Downy mildew	*Pseudoperonospora cubensis* (Berk. & Curt.) Rostow.
4.	Anthracnose	*Colletotrichum lagenarium* (Pass.) Ellis & Halsted [= C. *orbiculare* (Berk. & Mont.) Arx.]
5.	Fusarium wilt	*Fusarium oxysporum* Schlecht. emend. Snyder & Hansen, *F. o.* f.sp. *benincasae* Gerlagh & Ester , *F. o.* f.sp. *cucumerinum* Owen, *F. o.* f.sp. *lagenariae* Matuo & Yamamota, *F. o.* f.sp. *luffae* Kawai et al., *F. o.* f.sp. *melonis* Snyd. & Hansen, *F. o.* f.sp. *momordicae* Sun & Huang, *F. o.* f.sp. *niveum* (Smith) Snyder & Hansen
6.	Fruit and vine rot of pointed gourd	*Phytophthora melonis* Katsura
7.	*Alternaria* leaf blight	*Alternaria cucumerina* (Ell. & Ev.) Elliot
8.	*Pythium* fruit rot	*Pythium* spp., *Pythium butleri* (cottony leak) Subramaniam, *P. aphanidermatum* (Edson) Fitzp.

9.	Net blight/web blight/ leaf blight/belly rot	*Rhizoctonia solani* Kuhn
10.	Fusarium root rot (Crown and foot rot)	*Fusarium solani* f.sp. *cucurbitae* Snyder & Hansen.
11.	*Alternaria* fruit rot of pointed gourd	*Alternaria alternata* (Fr.) Kiessler
12.	*Rhizoctonia* fruit rot	*Rhizoctonia bataticola* (Taub.) Butler
13.	Stem rot/collar rot	*Sclerotium rolfsii* Sacc.
14.	*Rhizoctonia* root rot	*Rhizoctonia bataticola* (Taub.) Butler
15.	Charcoal rot of fruits	*Macrophomina phaseolina* (Tassi) Goid. [=*Macrophomina phaseoli* (Maubl.) Ashby]
16.	*Cercospora* leaf spot	*Cercospora* spp., *Cercospora citrullina* Cooke
17.	White mould	*Sclerotinia sclerotiorum* (Lib.) de Bary
18.	*Choanephora* rot	*Choanephora cucurbitarum* (Berk. & Ravenel) Thaxt.
19.	*Fusarium* fruit rot of pointed gourd	*Fusarium equiseti* (Corda) Sacc.
20.	Leaf spots	*Helminthosporium rostratum* Drechsler, *Phyllosticta cucurbitacearum* Sacc.
21.	Marginal leaf blight	*Exserohilum rostratum* (Drechsler) Leonard & Suggs
22.	Alternaria leaf spot	*Alternaria alternata* f.sp. *cucurbitae* Vakal.
23.	*Corynespora* blight/ target spot	*Corynespora cassiicola* (Berk and Curtis) Wei.
24.	Gray mold	*Botrytis cinerea* Pers.
25.	Collapse of melon	*Monosporascus eutypoides* (Petr.) Arx.
26.	Phoma blight	*Phoma exigua* var. *exigua* Sacc.
27.	Crater rot (fruit)	*Myrothecium roridum* Tode: Fr.
28.	*Phytophthora* root rot	*Phytophthora* spp., *Phytophthora capsici* Le.
29.	Black root rot	*Thielaviopsis basicola* (Berk. & Br.) Ferraris
30.	Verticillium wilt	*Verticillium albo-atrum* Reinke and Berthold.
31.	Sudden wilt	*Pythium aphanidermatum* (Edson) Fitzp.
32.	Gummy stem blight (vine decline)	*Didymella bryoniae* (Fuckel) Rehm (= *Mycosphaerella melonis* (Pass.) Chiu & Walker.
33.	Septoria leaf blight	*Septoria cucurbitacearum* Sacc.
34.	*Lasiodiplodia* vine decline/ fruit rot	*Lasiodiplodia theobromae* (Pat) Griffon & Maubl. (= *Diplodia natalensis* Pole Evans.

35.	Monosporascus root rot	*Monosporascus cannonballus* Pollock & Uecker.
36.	Net spot	*Leandria momordicae* Rangel
37.	Purple stem	*Diaporthe melonis* Beraha & O'Brien.
38.	Pink mold rot	*Trichothecium roseum* (Pers.) Link.
39.	Phomopsis black stem	*Phomopsis sclerotioides* Van Kesteren.
40.	Ulocladium leaf spot	*Ulocladium consortiale* (Thum.) Simmons.
41.	Scab/gummosis	*Cladosporium cucumerinum* Ellis & Arthur.
42.	Plectosporium blight	*Plectosporium tabacinum* (Beyma) Palm.
43.	Blue mold rot	*Penicillium* spp., *P.digitatum* (Pers.:Fr.) Sacc.
44.	*Myrothecium* canker (black canker)	*Myrothecium roridum* Tode.

Table 3. Post harvest diseases caused by the fungi and fungus-like organisms.

S.No.	Name of the disease	Causal organism
1.	White cottony rot	*Fusarium* spp.
2.	Soft rot	*Rhizopus* spp., *Mucor* spp., *Fusarium* spp., *Pythium butleri* Subram., *Choanephora* sp.
3.	Cottony leak	*Pythium aphanidermatum* (Edson) Fitzp. *P. butleri* Subramaniam and *P. cucurbitacearum* Takim.
4.	Calyx end rot	*Thielaviopsis paradoxa* (De Seynes) Hohn.
5.	Waxy rot or sour rot	*Geotrichum candidum* Link., *Fusarium* sp.,
6.	Anthracnose	*Colletotrichum capsici* (Syd.) E. J. & Bisby and *C. lagenarium* (Pass.) Ell. et Halst.
7.	Brown rot	*Botryodiplodia theobromae* Pat., *Aspergillus niger* van Tieghem, *A. flavus* Link, *Alternaria* spp., *Fusarium equiseti* (Corda) Sacc., *F. oxysporum* Schlecht. emend. Snyder & Hansen, *Macrophomina phaseolina* (Tassa) Goid.
8.	Stem end rot	*Botryodiplodia theobromae* Pat,
9.	Charcoal rot	*Macrophomina phaseolina* (Tassa) Goid.
10.	Green mould rot	*Aspergillus fumigates* Fres., *Aspergillus niger* van Ti.
11.	Gray white rot	*Sclerotium rolfsii* Sacc.
12.	Dirty grey rot	*Rhizoctonia solani* Kuhn.
13.	Black spot	*Curvularia ovoida* (Hiroe &Watan.) Munt.

Table 4. Diseases caused by bacteria and phytoplasma

S.No.	Name of the disease	Causal organism
1.	Angular leaf spot	*Pseudomonas syringae* pv. *lachrymans* (Smith & Bryan) Young *et al.*
2.	Bacterial wilt	*Erwinia tracheiphila* (Smith) Bergey *et al.*
3.	Bacterial leaf spot	*Xanthomonas campestris* pv. *cucurbitae* (Bryan) Dye
4.	Bacterial soft rot	*Erwinia carotovora* subsp. *carotovora* (Jones) Bergey *et al.*
5.	Brown spot	*Erwinia ananas* Serrano
6.	Bacterial rind necrosis	*Erwinia* spp.
7.	Bacterial fruit blotch/ seedling blight	*Acidovorax avenae* subsp. *citrulli* (Schaad *et al.*) Willems *et al.* (= *Pseudomonas pseudoalcaligenes* subsp. *citrulli*)
8.	Phyllody	Phytoplasma
9.	Witches' broom	Phytoplasma

Table 5. Diseases caused by the viruses

S.No.	Name of the disease	Name of virus
1.	Cucumber green mottle	Cucumber green mottle mosaic virus (CGMMV)
2.	Cucumber mosaic	Cucumber mosaic virus (CMV)
3.	Water melon mosaic	Water melon mosaic virus 1 and 2 (WMV 1 and 2)
4.	Zucchini yellows	Zucchini yellows mosaic virus (ZYMV)
5.	Chlorotic leaf spot	Bean yellow mosaic virus (BYMV)
6.	Mosaic	Potato virus Y (PVY) Papaya ring spot virus (PRSV-W)
7.	Cucumber latent	Cucumber latent virus (CLV)
8.	Tobacco ring spot	Tobacco ringspot virus
9.	Curly top	Beet curly top virus (BCTV)
10.	Cucumber vein yellowing	Cucumber vein yellowing virus (CVYV)
11.	Lettuce infectious yellows	Lettuce infectious yellows virus (LIYVV)
12.	Melon leaf curl	Melon leaf curl virus (MLCV)
13.	Melon necrotic spot	Melon necrotic spot virus (MNSV)
14.	Muskmelon vein necrosis	Muskmelon vein necrosis virus (MkVNV)
15.	Squash leaf curl	Squash leaf curl virus (SqLCV)
16.	Squash mosaic	Squash mosaic virus (SqMV)
17.	Tomato spotted wilt	Tomato spotted wilt virus (TSWV)

2.5.1 Fungal Diseases of Cucurbits

2.5.1.1 Seed rot and Damping-off

The disease is also known as seedling blight. A wide numbers of cucurbits are infested by this disease and cause substantial loss.

Symptoms

Damping-off of seedlings is a widespread disease caused by fungi that survive in soil, or the organisms associated with the seed and when the seeds are sown they attacked by the fungi, which survive in soil or seed coat. The causal pathogens attack seeds, seedlings and root of the cucurbit plants in fields and green houses. It causes maximum damage to the seeds before emergence (seed rot) and seedlings after emergence from the soil surface (damping-off). When comparatively older plants are attacked they are not always killed, but lesions are developed on the roots and stems, and in this case returded plant growth is observed (Chattopadhyay and Mustafee, 2008). Cucurbit seeds need warm soils to germinate and develop properly (18°C at 5cm depth). Seeds that are planted in cold, wet soils are at risk of seed rot and damping-off. These diseases are caused by several different fungi including *Pythium, Rhizoctonia, Fusarium etc.* Damping-off fungi infect and cause pre-emergence seed rot as well as post emergence rot of young seedlings. Infected seeds may not emerge from the soil due to decay of the seed immediately following seeding or germination. Seedlings may emerge with soft brown water soaked areas on the cotyledons (seed leaves). Stems may be thin, wire-like and unable to support even the small seedling. Fine cobweb like fungal mycelia may be visible on infected portion of seedlings. The fungi may attacks the stem at the soil line causing a constriction which causes the young seedling to topple over. This disease may also cause root rot. Often, plants that have survived damping-off might show symptoms of root rot. Roots can have a watery grey appearance, particularly the finer feeder roots. Infected plants are unlikely to grow into a mature plant.

The causal pathogens and perpetuation

The disease is caused by quite a few fungi which live usually in soil are *Pythium aphanidermatum, P. debaryanum, P. myriotylum, P. butleri, Rhizoctonia solani, R. bataticola, Phytophthora parasitica, Fusarium spp., Fusarium equiseti, Gibberella intricans* [teleomorph], *Acremonium* spp., *Thielaviopsis basicola* and some other fungi (Nagaich and Singh, 1960, Aulakh, 1971, Sinha, 2001, Chattopadhyay and Mustafee, 2008).

The most important damping-off causing fungal pathogen, *Pythium* spp. consists of much branched, hyaline, coenocytic mycelium. The hyphae penetrate the cell walls of the hypocotyl and ramify within and between the tissues of the cortical parenchyma. The sporangia and oospores are formed in the parenchymatic tissues of the host. Sporangia are spherical when terminal or oval or barrel shaped when intercalary. A prominent beak is formed on the sporangium, followed by the formation of a vesicle into which the sporangial contents are emptied. The protoplasm is divided to form reniform zoospores within the vesicle with two lateral flagella. The zoospores when released swarm for a few minutes and later form cyst. The cysts afterward germinate by producing the germ tubes. From the mycelium of the fungus oogonia and antheridia are formed. Oogonia are round shaped, and may be terminal or intercalary while antheridia are club-shaped. One or two antheridia are found attached to a single oogonium. The thick walled oospores are formed from fertilized oogonium, which afterward germinate by the formation of germ tube. The fungi multiply by zoospores as well as oospores. Sexual reproduction is isogamous (Chattopadhyay and Mustafee, 2008).

These fungi are common soil inhabitants. Some of them produce resting structures in soil. These fungi infect several weeds, and also survive on decaying plant material. They can be spread by infected transplant, water, movement of soil and other human activities.

Epidemiology

The severity of loss due to the disease depends upon the inoculum concentration of the causal organism in the soil and on the environmental conditions which predisposes the disease. An abundance of moisture in soil, high relative humidity and cloudy weather are especially favourable for the development of damping-off disease caused by *Pythium* spp. and other organisms in juvenile state of the plant.

Overcrowding of seedlings may increase susceptibility since it tends to delay cell wall thickening and lignifications (Nayak, 1998).

High soil moisture makes the soil nutrient available to the oospores which germinate to produce zoospores. It also helps in rapid spread of zoospores which attack the tiny seedling before their tissue hardening. The disease becomes more severe in poorly aerated and poorly drained soil. Such conditions are very common in heavy and compact soil.

The two fungi, *Phythium* spp. and *Rhizoctonia* spp. are commonly cause damping off in nursery beds. *Phythium* spp. primarily cause pre-emergence damping off at fairly cool temperature, whereas *Rhizoctonia* spp. cause late pre emergence and early post emergence damping off at slightly higher temperature. The disease appears in a wide range of temperature i.e. 10-35 C with 60-70% relative humidity, and the disease may cease at 4 C (Nayak, 1998).

Disease management

The disease usually causes sporadic outbreak and becomes difficult to manage.

a) Management of seedling diseases involved especial attention to the cultural practices. Nursery bed should be prepared at well drained field having preferably light soil. Well decomposed manures need to be applied in the nursery soil.

b) Summer plowing, and soil solarization using white, transparent polythene sheet at least 15-20 days is beneficial.

c) The disease may be checked by preparing the seed bed in such a place where sunlight is available throughout the day.

d) Fungicidal seed treatment with Thiram 75% WP, Copper oxychloride 50% WP and Mancozeb 75% WP at 3-4g/kg seed or bio-inoculation of seed with *Trichoderma viride* and *Azotobacter chroococum* provide to some extent of protection against damping-off (Nath, 2004, AICVIP, 1997-98). Seed treatment with *Trichoderma viride* formulation @ 4g + Metalaxyl 35% @ 6 g per kg seed is also recommended (Gour *et al.*, 2008).

e) Soil inoculation with bio-antagonists before sowing is efficacious.

f) Light but frequent irrigation of nursery bed helps in reducing the incidence of seedling diseases (Nayak, 1998). Careful irrigation management is important. Drip irrigation is more effective than other method of irrigation.

g) Care should be taken during transplanting of cucurbit seedlings when seedlings prepared in plastic tray or pot. Disturbance of the plants root system or damage of the seedling at collar region can make congenial condition for the fungi.

h) Soil drenching with 1% Bordeaux mixture or 0.2% Captan 75% WP or 0.3% Copper oxychloride 50% WP can help to prevent the serious disease (Gour *et al.*, 2008).

i) Drenching of seedlings with appropriate fungicides can reduce the disease severity. Spraying of systemic fungicides like Metalaxyl 35% WS @ 0.1% or combination of contact and systemic fungicides like Carbendazim 12% + Mancozeb 63% WP @ 0.25% is quite helpful.

2.5.1.2 Powdery mildew

Powdery mildew is one of the most common diseases of cucurbits. It is known to occur on the cucurbits since the year 1800. Six species of powdery mildew fungi have been reported worldwide on various hosts. Among them, occurrence of *Erysiphe cichoracearum* DC. and *Sphaerotheca fuliginea* (Schl.) Salmon *[also known as Podosphaera xanthii* (Castagne) U. Braun & Shishkoff (Seebold, 2010)] are very common and widespread in distribution. In Asia, the disease is reported to occur in Iran, Iraq, Saudi Arabia, Isreal, Malaysia, Singapore, China, Japan, Taiwan and India. In India, the disease was reported first time by Butler in 1918 on various hosts from Uttar Pradesh and Bihar. He identified it on the basis of conidial characters (Sharma, 2005). The disease is also reported in Malaysia and Singapore on pumpkin (Thompson, 1933). Jhooty (1967) observed that the disease was caused by *Sphaerotheca fuliginea* in Punjab. The disease is prevalent in several states of India viz. Haryana, Punjab, West Bengal, Tamil Nadu, Uttar Pradesh, Jammu and Kashmir, Assam, Karnataka and Maharashtra (Mustafee, 1998). It occurs most severely during dry seasons than the wet weather condition (Gour *et al.*, 2008). The disease caused by *Erysiphe cichoracearum* DC. has been reported on *Momordica balsamina, Trichosanthes dioica* (Butler and Bisby, 1931), *Lagenaria vulgaris* (Rajendran, 1965), *Coccinia cordifolia, Coccinia indica* (Khan *et al.*, 1972, Sohi *et al.*, 1981), *Benincasa hispida* (Khan *et al.*, 1972). Although the fungus attacks all the cucurbits it is especially destructive on pumpkin (*Cucurbita moschata*) and bottle gourd (*Lagenaria siceraria*). The bitter gourd (*Momordica charantia*) is the least affected crop. Both the powdery mildew causing fungi occur commonly on cucumber also.

Symptoms

The disease is usually found on the foliage and green stems (vines), and rarely on fruits. It is characterized by the appearance of minute, white to dirty grey spots (sometimes with a reddish brown tinge) mostly on the upper surface of leaves, also found on the lower surface and vines. On older leaves or on the shaded portion of plant the disease appears first. As the increase in size, the spots become powdery (talcum like growth), consisting of superficial mycelium, conidiophores and conidia. The superficial powdery mass may ultimately cover the entire affected green surface of the host. The affected leaves turn yellow and become necrotic. Affected leaves and stem may dry-off in case of severe infection and further growth of the plant is arrested, which leads to premature death of the vines. While powdery mildew primarily infects leaves and vines, infections rarely take place on pointed gourd, cucumber or melon fruit. Squash fruit are not directly infected. Regardless of direct infection of the fruit, infected plants produce fewer and smaller fruit(s). The fruits remain undersized and sometimes are deformed. The yield and quality of the fruit is reduced greatly. Infected plants produce fruits with incomplete ripening, poor storability and poor flavor. In India, mostly in the winter season crops, minute, dark brown to black pin-head bodies appear rarely intermixed with the white powdery mass. These are the cleistothecia of the fungus. Powdery mildew is most severe after fruit-set and in densely planted fields.

The causal pathogens and perpetuation

Cleistothecia of six species of Erysiphaceae were recorded on cucurbits in different parts of the world. *Erysiphe cichoracearum* DC. and *Sphaerotheca fuliginea* (Schl.) Salmon are two most regular and most commonly distributed species all over the world. There are two situations viz. two pathogen disease situation and one pathogen disease situation prevails throughout the world on the basis of identity of causal organism (Sharma, 2005).

Two pathogen disease situations refers that the same disease is caused by two pathogens separately. This situation should not be confused with diseases caused by two pathogens in combination like disease complexes, which involves the intimate association of two or more pathogens. Two pathogen disease situations have been established in U.S.A., U.K., Germany, Italy, Hungary, Bulgaria, ex-U.S.S.R., Japan, Israel and India. *Erysiphe cichoracearum* dominates over *Sphaerotheca fuliginea* in U.S.A., U.K., Germany and Bulgaria, while *Sphaerotheca fuliginea* is predominant in Italy, Japan, Israel and India. In Bulgaria and ex-U.S.S.R., both are more or less equally important (Sharma, 2005).

One pathogen disease situation by *Erysiphe* is found in Canada, France, Norway, Sweden, Austria, Switzerland, Mozambique, Egypt, Malta, Fizi, Kenya, Bolivia, Brazil, Peru, Nicaragua, West Indies, Iraq, Saudi Arabia, Malaysia and Singapore. One pathogen disease situation by *Sphaerotheca* is recognized in Natherlands, Greece, Turkey, Czeckoslovakia, Rumania, Australia, Newzealand, South Africa, Sudan, Malawi, Iran, China and Taiwan (Sharma, 2005).

In India, the disease is caused by *Erysiphe cichoracearum* as well as *Sphaerotheca fuliginea* (Sharma, 2005).

Both the fungi produces two spore types: the white powdery spores present on the plant surface are conidia and those produced in tiny round fruiting bodies (Cleistothecia) are ascospores.

Erysiphe cichoracearum is known to produce the oidial or conidial stage as well as the perfect stage. The fungus produces well developed superficial, evanescent but sometimes persistent and effused mycelium with well developed haustoria in host cells. The mycelium is spread on the leaf surface and the hyphae draw nutrients by sending haustoria into the epidermal cells of the host plant. Unbranched, erect conidiophores arise from mycelial web that produces single celled, hyaline, ellipsoidal or barrel-shaped conidia (25-45 × 14-26 μm) in chains at their apex abundantly and disseminated mainly by wind. The conidiophores are almost at right angles to the host surface. The fungus is probably heterothallic. The fungus produces thick-walled, dark coloured, gregarious or scattered globose cleistothecia (90-135 μm in diameter) that contain ovate to broadly ovate, rarely subglobose, and more or less stalked asci (60-90 × 25-50 μm). Each cleistothecium contains 10-25 numbers of asci. Usually two, rarely three ascospores formed in each oval or sub-cylindrical ascus. Ascospores are hyaline, elliptic and thin walled measuring 20-30 × 12-20 μm. Appendages are numerous, myceloid, basally inserted, hyaline to dark, interwoven with mycelium, 1-4 times as long as the diameter of cleistothecium, rarely branched (Gupta *et al.*, 2001, Khatua and Saha, 2004, Rangaswami and Mahadevan, 2004, Sharma, 2005).

Sphaerotheca fuliginea causes serious infection in gourd, squash, pumpkin, cucumber and melons. Sexual stage has been reported on *Cucurbita moschata* from Himachal Pradesh (Sohi and Nayar, 1969), *Lagenaria leucantha* and cucumber from Uttar Pradesh (Khan *et al.*, 1976). *Erysiphe cichoracearum* and *Sphaerotheca fuliginea* occurs commonly on cucumber also (Mital and Akram, 1985). The conidial stage has been noticed on *Lagenaria siceraria, Luffa aegyptiaca, Cucumis melo, Cucumis melo* var. *utilissimus* (Jhooty, 1967) and many other cucurbits. When the pathogen inoculated artificially, a large number of cucurbits have been found susceptible in Uttar Pradesh, Himachal Pradesh and Karnataka (Khan *et al.*, 1972, Sohi, 1984). The fungus produces intercellular, hyaline mycelium (rarely brown when old) usually evanescent but sometimes persistent forming white circular to irregular patches on the host surface, with haustoria. At the tip of the conidiophores (short and simple) the fungus produces long chains of conidia that are ellipsoid to barrel shaped and 25-37 × 14-25 μm in size. Cleistothecia produced by the fungus are scattered to densely gregarious, 66-98 μm in diameter. Appendages are variable in number, usually as long as the diameter of cleistothecium, myceloid, brown, tortuous, interwoven with mycelium. Each cleistothecium contains single broadly elliptic to subglobose ascus with 50-80 × 30-60 μm in size. Ascospores are ellipsoid to nearly spherical, 17-22 × 12-20 μm; each ascus contains eight numbers of ascospores (Sharma,

2005). On the basis of cross inoculation method, Kaur and Jhooty (1985) reported three distinct types of isolates of the fungus (*S. fuliginea*) on different cucurbits. On the basis of four differential hosts (viz. PMR-45, PMR-59, PMR-6 and Edisto-47), they also observed the widespread occurrence of race-3 of the fungus in Punjab (Kaur and Jhooty, 1986).

Complete disease cycle is not well understood. The fungus produces conidial and perfect stages. Probably, the fungus survives in conidial stage on wild cucurbitaceous hosts and other collateral hosts and the wind-borne conidia cause secondary spread. The conidia are capable of germinating even under dry condition with low humidity and therefore the secondary infection occurs very rapidly. With maturity of the host plant and as the disease advances, the fungus produces sexual organs, resulting in the formation of the cleistothecia containing ascospore, which enable the fungus to survive in the field off-season, when susceptible host plants are grown, the ascospores germinate to cause fresh infection. Since the fungus has a wide host range, it can also perpetuate in its conidial stage, multiplying on the collateral hosts and becoming virulent when the climatic conditions are optimum and the host plants are vulnerable to infection (Rangaswami and Mahadevan, 2004, Sharma, 2005).

In India, cleistothecia of both *Erysiphe* and *Sphaerotheca* were reported on different cucurbits from field as well as greenhouse condition. Cleistothecia are produced in nature only during late winter or early summer whereas cleistothecia of *Sphaerotheca* in greenhouse condition develop during December to January. Temperature and host physiology play an important role in the production of cleistothecia (Sharma, 2005).

There are some reports regarding the production of perithecia in nature on some cucurbits in Indo-Gangetic plains, but their role in perpetuation of disease seems to be doubtful, because of the existing adverse weather condition during summer in this region. Chance of oversummering of cleistothecia due to intense high temperature during summer and heavy rains during rainy season appears to be remote (Sharma, 2005).

The pathogen overwinters through various means. The possibility of overwintering or oversummering of the fungus as mycelium in buds is excluded as the majority of the cucurbits are cultivated as annual crops. In subtropical and tropical parts, overwintering may not be a problem and the fungus may survive well as cleistothecia on a number of congenial hosts mainly gourds climbing on the house hut tops during winter in rural India. Cucurbits are grown in fields or kitchen gardens as vegetable crops all over the year in India. However, the fungus may not be able to establish on them due to intense heat of summer followed by heavy rains. The cucurbits are grown as key vegetable crops during summer in hilly tracts of Himalayas. The fungus may oversummers on these summer crops in the hills and blowing each year to the plains during winter. The inoculum during summer may also be blown from plains to hills. Moreover, some shade plants provide protection to some cucurbits, and the fungus may survive on these cucurbits during summer in plains itself (Sharma, 2005). In Punjab, Jhooty (1967) observed that the vegetative mycelium and conidia are maintained on sheltered cucurbits and these are important in disease cycle. Some collateral or alternate hosts most likely play the same role in the annual recurrence of the disease, as the pathogen has a wide host range. Though, the mode of perpetuation and recurrence of the disease are still unsolved (Sharma, 2005), the contradictory statement has also been reported by Gour *et al.* (2008). They stated that the main source of primary inoculum is wild and cultivated cucurbits in one or the other locality from where the conidia are blown to the new crop. Mustafee (1998) also reported that the mode of survival of the pathogen is soil and air.

Epidemiology

Unlike many other fungal diseases that need leaf wetness for infection, moisture on plant surfaces actually inhibits the powdery mildew fungus. High humidity, however, is required for infection. Older leaves are more susceptible. Densely planted crops, plants crowded by weeds, plants in shaded sites, and over fertilized cucurbit plants are more likely to be infected with the disease. The disease is also favoured by sultry weather, moderate temperature and reduced light intensity. For good conidial germination, a temperature range of 22-31°C with optimum of 28°C, and low RH of 20 per cent for less than 2 hours is essential (Sohi, 1984).

Disease management

a) The wild cucurbits should not be allowed to grow near the cultivated cucurbitaceous crop field (Gour *et al.*, 2008).

b) Search for disease resistance in host plants is most important. The following resistant sources are important to avoid the severity of the disease. For melon PMR 45, PMR 450, PI 124111, and for cucumber PI 200815, PI 200818, *C. hardwikkii,* wise 2757 (USA) are very much important for researchers to develop resistant varieties. Many researchers have screened germplasm to identified sources of resistance to the disease (Swamy *et al.*, 1981, Amin *et al.*, 1982). As many as 31 Indian and 10 exotic germplasm of musk melon reported to be highly resistant to the disease (Swamy *et al.*, 1981). Waraitch *et al.* (1977) reported some resistant varieties of musk melon namely PMR 5, PMR 6 and Arka Rajhans. Other reported important resistant varieties of musk melon are Diguria and Haragola (Amin *et al.*, 1982). Arka Manik is one of the resistant varieties of water melon.

c) Fertilizer should be applied based on soil test results. Over doses of nitrogenous fertilizers may increase the disease severity.

d) Good air movement around the crop microclimate through proper spacing, staking of plants and weed control is very much important.

e) Seed treatment and soil drenching with systemic fungicides may protect the young seedlings from the disease.

f) Destruction of diseased plants by burninmg, especially late in the off-season is important to avoid the formation of sexual stage of the pathogens.

g) Several phylloplane fungi including *Penicillium fellutanum* have been reported to inhibit spore germination of *E. cichoracearum* on *Cucurbita maxima* that can be used as bio-antagonists (Srivastava and Suman, 1986).

g) Fungicides should be applied when a single spot of powdery mildew is first found. In earlier years, this disease used to be controlled by sulphur dusting, but most of the cucurbits are susceptible to sulphur injury, especially when it is done during the hot days. Fortnightly sprays of fungicides like Dinocap 48% EC (0.05 – 0.1%), Carbendazim 50% WP (0.1%), Thiophanate methyl 50% WP (0.1%), Tridemorph 80% EC (0.05%), Benomyl 50% WP (0.1%) or Wettable sulphur (0.2%) etc. have been found useful in controlling this disease on bottle gourd, summer squash and pumpkin (Suhag and Mehta, 1982, Choudhury, 1990, Chadha, 2001, Khatua and Saha, 2004, Annonymous, 2004). Dusting with fine sulphur to cover the foliage @ 17 kg/ha, once or twice during the period of plant growth can manage the disease effectively (Roychaudhury and Verma, 1975a, Rangaswami and Mahadevan, 2004). The disease can be managed effectively, if these fungicides sprayed when the first initial symptoms occur. Copper oxychloride 50% WP (0.4%) or Mancozeb 75% WP (0.2%) has also been recommended to spray at 8 days interval. About 3-4 sprays will be enough to manage the disease. About 2-3 sprayings of Sulphur 80% WP at 0.2% can also be used at an interval of 5-6 days (Sinha, 1990). Spraying of Tridemorph (0.05%) or Sulphur 80% WP (0.25%) at 15 days intervals starting from 30-40 days after sowing has been recommended to control powdery mildew of melon under Rahuri conditions (Rai *et al.*, 2008a). Spraying with Sulphur 80 WDG/80 WP, Carbendazim 50 WP, Myclobutanil 10 WP, Triadimefon 25 WP and Flusilazole 40 EC at the recommended dosages offers good control of powdery mildew (Annonymous, 2010). As per recommendation of Mustaffe (1998), Bhattachatya *et al.* (2006a) and Gour *et al.* (2008) spraying of Dinocap 48% EC (0.05%), Carbendazim 50% WP (0.1%), Benomyl 50% WP (0.1%) and Tridemorph 80 EC (0.05%) are much effective.

2.5.1.3 Downy mildew

Downy mildew is an important fast-spreading fungal disease affecting most of the cucurbits like sponge gourd, ridge gourd, ash gourd, pointed gourd, muskmelon, long melon, cucumber, bottle gourd, bitter gourd, snake gourd, vegetable marrow, pumpkin and *Cucumis callosus* etc. It is, though, more severe and widespread on muskmelon, sponge gourd, ridge gourd, squash and cucumber (Sohi and Sharma, 1998). The disease is fairly common in northern India, where it becomes serious during latter part of the rainy season. The disease was first reported and described from Cuba in 1868 by Barkeley and Curtis (Walker, 1952). The disease was also reported to occur on pointed gourd from Bihar and West Bengal (Khatua *et al.*, 1981a, Mondal *et al.*, 2014). Downy mildew is especially damaging in worm and humid climates where the pathogen thrives well.

Symptoms

Downy mildew symptoms are found almost exclusively on leaves. Lesions vary considerably but usually appear first as small, slightly chlorotic to bright yellow areas on the upper leaf surface in the crown area of the plant resemble those of mosaic mottling. The pale green areas are separated by islands of darker green. One or more spot may be formed on a single leaf. As lesions expand and number of lesions increases, they may become necrotic and brown, eventually coalescing and killing the entire leaf. Heavy infection will lead to the browning of leaves and eventual death of the entire vine, and plants will appear scorched if cool and wet weather persists. Usually, the central leaves are attacked first and are followed by other leaves until entire plant is wilted or weakened. Young leaves are less susceptible than older ones. Infection occurs more readily on lower surface than on upper surface. Premature defoliation caused by downy mildew will reduce fruit size and predispose fruit to sunscald injury.

Margins of the lesions are usually irregular, sometimes 'blocky' or angular in appearance and tend to be limited by leaf veins that can be confused with the bacterial disease angular leaf spot. This type of symptom is most distinct in cucumber. On the lower surfaces of the leaves, lesions will be water-soaked and slightly sunken, and are less vivid in color. In moist weather (high humid) condition, downy mildew will only sporulate profusely on the underside (bottom lesions) of leaves producing light to dark gray or purplish-brown spores in white, fuzzy fungal growth (downy appearance). But on pointed gourd, downy growth of the fungus is not found even in humid weather or rainy day (Mondal *et al.,* 2014). On watermelons, an exaggerated upward leaf curling is very common. Lesions are usually small on snake gourd and ash gourd.

The causal pathogen and perpetuation

Downy mildew is caused by the fungus-like organism, *Pseudoperonospora cubensis* (Berk. & Curt.) Rostow., is an oomycete that is not a true fungus and is often referred to as a water mold due to the fact that it thrives in wet or very humid conditions.

Mycelium of the pathogen is hyaline, coenocytic and intercellular, develops abundantly in the mesophyll, but also penetrates palisade tissues. Haustoria are small, ovate, intercellular sometimes with finger like branches. Sporangiophores are 180-400μm in length, dichotomously branches in their upper third, emerge in groups of 1-5 through stomata. The soporiferous tips, on which the sporangia are borne singly are subacute. Sporangia are pale grayish to olivaceous purple, ovoid to ellipsoidal, thin-walled, with a papilla at the distal end, measuring 20-40 × 14-25 μm. Sporangia germinate by production of flagellate zoospores, rarely by infection hyphae. Oospores are not common in the species but when produced are thick walled, smooth light yellow (Gupta *et al.,* 2001, Khatua and Saha, 2004, Rangaswami and Mahadevan, 2004, Sharma, 2005, Mondal *et al.,* 2014).

The pathogen survives both as mycelium and sporangium from one season to another, since oospores are not common. Sporangia survive in cold weather (Khatua and Saha, 2004). Cucurbit crops including wild species affected by downy mildew and old infected crop trash are the primary source of inoculum. Airborne sporangia from these infected cucurbits cause primary infection. Secondary infection within a field may be spread mostly by air currents, as well as rain splash, workers and agricultural implements. The sporangia can also be carried by cucumber beetles (Lange *et al.,* 1989). After release, the zoospores on germination produce germ tube or germinate directly and the germ tube penetrates the host surface and cause infection (Khatua and Saha, 2004). The incubation period is four to 12 days, depending on the temperature and photoperiod. Oospores are not generally observed on the host plants. But, Khosla *et al.* (1973) reported oospores on pointed gourd in Madhya Pradesh and Bains *et al.* (1977) on *Melothrina maderaspatana* in Punjab. Perpetuation of the fungus in the form of active mycelium on self-sown or cultivated sponge gourd plants growing in sheltered places were observed by Bains and Jhooty (1976a) from Punjab at Ludhiana during severe winter, and they recorded sporangial production from November to February. In south Indian condition, the downy mildew fungus perpetuates on various cucurbit hosts without facing any difficulty due to monoculture of the crop as well as prevailing milder climate. At Solan, Himachal Pradesh, the disease is only noticed on snake gourd and ash gourd without forming oospores (Sohi and Sharma, 1998).

Bains and Jhooty (1967b) observed that the organism causing downy mildew on muskmelon did not infect ash gourd and pumpkin, whereas the reverse was possible. Bains and Sharma (1986) have proposed the existence of races on the basis of different rates of disease spread and the reaction of different isolates of pathogen on 18 cucurbit hosts. Bains and Sharma (1986) reported the occurrence of two new races from Punjab.

Epidemiology

Infection of the pathogen generally occurs through stomata. The disease is favored by prolonged periods of cool, wet weather. Moist conditions, such as rain, fog and heavy dews required for successful development of the disease. Disease is more common in areas with high rainfall. The incubation period of the disease is one week at 22-28°C. The pathogen can infect the plants at temperature between 10-27°C, with optimum day temperature of 25-30°C and night temperature of 15-21°C. The optimum temperature for infection is 20°C. However, the temperature above 35°C arrested the further invasion of the host. Relative humidity of more than 75 per cent favour the disease development. For sporulation (i.e. for production of sporangia), at least 6 hours of wet period (100 per cent relative humidity) at the leaf surface and temperatures of 5-30°C after 6 hours of dry period is essential.

At 15-20°C temperature, optimum sporangial production takes place. Germination of sporangia requires free moisture and started within 1 hour and reaches the maximum in 2 hours at 20°C. Maximum dispersal of sporangia occur between 6 to 10 a.m. Planospores remain motile for longer period at 10 and 15°C than 30 and 35°C. A film of moisture on leaf surface is necessary for the infection to occur (Lange *et al.*, 1989, Khatua and Saha, 2004). Khatua and Saha (2004) reported that the downy mildew of pointed gourd usually appears in December and become severe in February-March. Spread and development of the disease is arrested after middle of April when temperature become hot and humidity become low (Mondal *et al.,* 2014).

Disease management

The disease is very difficult to manage once it gets started.

a) Monitoring of crops closely to identify the disease early in its cycle and should be alert to weather conditions that may cause an early infection.

b) Affected vines or leaves should be removed from the field.

c) Wild cucurbits from vegetable growing areas should be destroyed as they serve as secondary host.

d) It is better to avoid overhead irrigation. Drip irrigation and wide row spacing to promote leaf drying and encourage good air movement around the plants is helpful to manage the disease.

e) Seeds should be sown in sunny sites with good airflow.

f) Spraying of preventive fungicides can be recommended for management of this disease. Mancozeb 75% WP (0.2-0.25%), Chlorothalonil 75% WP (0.2%), Zineb 75% WP (0.2%) and Copper oxychloride (0.4%) at 8-10 days interval gives good control of the disease as protectant but failed to check established infection (Chadha, 2001, Gour *et al.,* 2008). Khatua *et al.* (1981a) reported the effectiveness of Metalaxyl for checking this disease on cucumber. Spraying of Mancozeb 75% WP (0.1%) and Metalaxyl 35% WS (0.05%) mixture starting with the onset of favourable weather at 10 days interval are very effective. Cymoxanil 8% + Mancozeb 64% and Chlorothalonil 75% WP as tank mixture also provide effective control for downy mildew of cucumber (Robak, 1995). Spraying the field with 2:2:50 Bordeaux mixture is effective (Choudhury, 1990, Roychaudhury and Verma, 2000). Dusting with tribasic Copper sulphate (5% copper content) @ 20-25 kg/ha early in the season, and 45-60 kg/ha when the vines are larger is recommended to manage the disease (Choudhury, 1990). Metalaxyl 8% + Mancozeb 64% WP (0.3%), Copper oxychloride 50% WP (0.4%) and Mancozeb 75% WP (0.25%) is very much effective under field condition. Retarded growth of vine of pointed gourd was reported due to spraying of hexaconazole and propiconazole (Saha, 2002, Khatua and Saha, 2004, Mondal *et al.*, 2014). Use a fungicide program that allows for the rotation of protective and systemic fungicides which reduce the chance of fungicide resistance developing. Spraying of Mancozeb (0.3%) at 15 days intervals has been recommended to control downy mildew of melon under Ludhiana condition. Mancozeb 75% WP, Zineb 75% WP or Fosetyl-AL 80% WP (0.3%) at 10 days intervals was recommended for management of the disease of cucumber under Bangalore conditions (Rai *et al.*, 2008). Spraying of Metalaxyl 8% + Mancozeb 64% WP and Copper oxychloride 50% WP at the recommended dosages reported effective (Bhattachatya *et al.*, 2006, Annonymous, 2010). Mancozeb 75% WP (0.3%) and Fosetyl-AL 80% WP (0.2%) is recommended by Mustafee (1998) for managing the disease effectively.

g) Resistant cultivars should be preferred for growing and avoiding the disease. Wild melon, *Cucumis callosus* is resistant to this disease that can be used as source of resistance by the researchers. In Georgia, MM-7 and Mathuria are reported as moderately resistant cultivar. The germplasm showed resistance to both downy and powdery mildew are IHR 142, IHR 157, IHR 180, H 190, H 226 and H 240 (Sohi and Sharma, 1998). Some other important resistant varieties of cucumber are Palmetto, PR 27, Santee and Palomar (Choudhury, 1990).

2.5.1.4 Anthracnose

Anthracnose was first described on gourd in Italy in 1867. In India, the disease has been reported from Punjab on long melon and gourd (Mundkur, 1937). The disease has also been observed on several other cucurbit hosts. Prakash *et al.* (1974) reported that the cultivation of water melon, musk melon, snake gourd, round melon, cucumber, ash gourd and bottle gourd has been limited to a great extent. The disease is reported from almost all cucurbits growing countries. It is very serious in watermelon, muskmelon, bottle gourd, cucumber, snake gourd etc. while squash and pumpkin are less infected cucurbits. Havoc losses occur on fruits of watermelon grown in temperate countries. As much as 90 per cent yield losses takes place in bottle gourd and water melon during rainy season as reported by Sohi (1975) from India. About 69-70 per cent of fruits show infection in musk melon, water melon and bottle gourd (Sohi and Sharma, 1998). Amin and Ullasa (1981) have recorded 63 per cent yield loss in water melon. Ullasa and Amin (1986) recoded as high as 99 per cent loss of marketable fruits. Losses in storage or during transport can occur when freshly harvested fruit becomes infected.

Symptoms

Symptoms of the disease vary according to the host. All above-ground plant parts are affected and plants can become infected at any stage in their development. The spots on the older leaves (foliage) begin as small yellowish or water soaked areas that enlarge rapidly and turn reddish brown to tan in most cucurbits but dark brown to black on the watermelon. Cucumber leaf spots are comparatively bigger than the watermelon leaf spot and often have a yellow halo. Spots are often circular to angular. Afterward, spots may coalesce to create extensive blighting. These areas become dry and tear away, typically giving the foliage a ragged appearance. Often the leaves at the center of a plant are attacked first, leaving the stem and runners bare. Tan to black, elongated, slightly sunken streaks (cankers) formed on petioles and stems that can girdle the vine, causing death of the tissue beyond the lesion. On cucumber (*kheera*) leaves, the spots commonly start on a vein and expand into brown spots which are angular or roughly circular. Growing leaves may be distorted and coalescing spots may cause death of the entire leaf. On muskmelon, the petioles are attacked so that often defoliation occurs. On pumpkin, symptoms include small, tannish-brown spots on the upper leaf surface that become pinkish-orange in colour. Shape of the lesions may be circular or spindle-shaped, which are limited to the veins of leaves. Older lesions eventually turn brown and often fall out leaving a 'shot holes' on infected areas of the leaf. Saha (2002) recorded the disease in the month of August on pointed gourd grown on scaffold. Individual vines are dry up due to infection on the stem. Stem lesions are brown in colour and 2-4cm in length without superficial mycelia growth.

Fruits, if infected early, may turn black, shrivel and die. This disease is most characteristic on fruits reaching maturity. On such fruits, the spots are roughly circular, water soaked with variable in size (0.6 to 6.5 cm in diameter) depending upon the age of the plant and weather condition. Spots turn dark green to brown in colour with age and may become sunken. Lesions may also develop on storage fruits. Fruit lesions on watermelon can be cracked and irregularly shaped.

On pumpkin, lesions are mostly circular, sunken, and measure about 0.2 to 0.5 cm in diameter or larger. On butternut squash the lesions are similar but may be larger and more elongate. Under humid conditions, the centre of these spot darkens and develops tiny black specks where salmon-pink masses of spores may be seen. Reddish gummy exudate may appear on spots. Fruit symptoms may also develop during transit.

The causal pathogen and perpetuation

This fungal disease is caused by *Colletotrichum lagenarium* (Pass.) Ellis & Halsted. A synonym for *Colletotrichum lagenarium* is *C. orbiculare* (Berk. & Mont.) Arx. The sexual or perfect stage (teleomorph) is *Glomerella lagenarium. C. lagenarium* can infect a plant successfully at any stage of growth. When moisture is present, the spore germinates and a penetration tube enters the plant within three days. Invaded tissues die and form a canker followed by the production of spores which are then ready to repeat the infection process (Ferreira and Boley, 1992).

Colletotrichum produces conidia in acervuli (the canker). Masses of conidia attain a pink or salmon color. Conidia are released from the acervuli and come into contact with susceptible plant and/or fruit hosts and germinate when water is present and temperatures are optimal (20-32 °C). The conidia germinate and penetrate host tissues directly (Ferreira and Boley, 1992).

Saha (2002) recorded the organism infecting pointed gourd is *Colletotrichum capsici* (Syd.) Butler. The organism, *Colletotrichum capsici* forms acervuli. Acervuli on stems are rounded or elongated, intra and sub-epidermal, disrupting outer epidermal cell walls of host. Setae are brown, 1-5 septate, rigid, hardly swollen at the base, slightly tapered to the paler acute apex, up to 250μm in length. Conidia are hyaline, falcate with acute apex and narrow truncate base, aseptate, 16-30 × 2.5-4.0μm, formed from unicellular hyaline to faintly brown cylindrical conidiophores (Khatua and Saha, 2004).

The disease is soil borne and also seed borne if fruits are attacked. Weed host belonging to cucurbitaceous family also serve as a source of perennation of the fungus (Gour *et al.*, 2008).

Epidemiology

The disease causing fungus is primarily soil-borne but it may also be seed borne if fruits are attacked and the fungal mycelium reaches the seed. The fungus overwinters in old cucurbit vines, in seed, or in weeds in the cucurbit family. The fungi produces spores (conidia) on infected leaves and fruit. These spores are easily spread by splashing rain, irrigation, activities of workers, agricultural equipments and insect activities. This soil-borne fungus is splashed with soil particles onto healthy leaves during rainfall or overhead irrigation.

The fungus occurs in epidemic form only when there is more than average rainfall. The pathogen can tolerate temperatures up to 45°C. The pathogen unable to infect the host below the temperature of 18°C. Warm temperatures (20 to 30°C), high humidity (frequent rains, poor drainage and 100% relative humidity for at least 18 hours) and high rainfall favour rapid development and spread the disease. Temperatures between 20 and 30^{0}C are most conducive for infection with day and night temperatures ranges of 26-30^{0}C and 18-20^{0}C respectively is most congenial for disease development (Sohi and Sharma, 1998). Intermittent rains during August and September coupled with warm temperatures cause epidemics in water melon, musk melon, and bottle gourd. Disease development is not affected by plant population (Ullasa and Amin, 1986). The disease is easily spread by splashing rain, irrigation water, on workers hands or equipment. In high humid condition the disease and its further spread is very much difficult to check. The conidia (spores) are released and spread only when the acervuli are wet and are generally spread by splashing water and blowing rain or by coming into contact with insects, other animals (including human) and tools. The fungus can overwinter as mycelium on or in seed and on residue from diseased plants in and on the soil. The fungus can also live in weeds of the cucurbit family (Ferreira and Boley, 1992). Anthracnose can appear anytime during the season, but most damage occurs late in the season after the fruit is set. At least three races of *Colletotrichum* have been reported.

Disease management

a) The disease can be managed by sowing certified, disease-free seed of a variety resistant to the races of *Colletotrichum* (watermelon - races 1 and 3; cucumber - races 1, 2, and 3) (Ferreira and Boley, 1992).

b) Cultivation should be done in well drained soil free from surface run-off water.

c) Three to four year crop rotation programme with non-host crop should be practiced (Kennelly, 2012).

d) It is better to plough down the field after harvest the crop. Plant debris should be collected for destruction by burning at the end of the growing season.

e) Drip irrigation can be advised instead of overhead sprinklers, if possible.

f) Removal of weeds under cucurbits family and volunteer cucurbits plants is beneficial.

g) Continuous cropping of different cucurbit vegetables in tropical regions especially in riverbed areas should be discouraged (Sohi and Sharma, 1998).

h) Disease incidence can be minimized by immunizing plants to some bacterial, viral, and fungal plant pathogens by exposing them to *C. lagenarium* (Ferreira and Boley, 1992).

i) The seeds should be treated with fungicides like Thirum 75% WP @ 2.5g/kg, Benlate 50% WP @ 1.0g/kg or Carbendazim 50% WP @ 1.0g/kg of seeds or bioantagonist like *Trichoderma viride* @ 6g/kg of seeds (Gour *et al.*, 2008).

j) The disease can be managed by the repeated spraying of recommended fungicides like Benomyl 50% WP, Carbendazim 50% WP and Thiophanate methyl 50% WP @ 0.1% or Mancozeb 75% WP @ 0.25%. (Amin and Ullasa, 1981, Ullasa and Amin, 1986, Mustafee, 1998, Chadha, 2001, Annonymous, 2004) or Mancozeb 75% WP, Zineb 75% WP, Captan 50% WP (0.2%) etc. at an interval of 5-7 days. Khatua and Saha (2004) stated that one spray with Mancozeb 75% WP (0.2%) or Carbendazim 50% WP (0.1%) after removal of infected vines will check the disease. Chlorothalonil 75% WP and Maneb are also efficacious against the disease (Zitter, 1987). Spraying of Carbendazim 50% WP and Benomyl 50% WP @ 0.2% at 15 days intervals has been recommended for controlling of the disease of watermelon under Bangalore condition (Rai *et al.*, 2008). Gour *et al.* (2008) also recommended Copper oxychloride 50% WP @ 0.3% for controlling the disease.

2.5.1.5 Fusarium wilt

Fusarium wilt, caused by soil-borne fungus of the genus *Fusarium*, affects most cucurbits like watermelon, muskmelon, bottle gourd etc. Several Fusarium species and physiological races have been identified on cucurbits from different parts of the world.

Fusarium wilt of melon (*Fusarium oxysporum* f.sp. *melonis*) was first reported as a disease of melon in the United States in New York in 1930, although pathogenicity was not confirmed until the disease was described in Minnesota in 1933. Although the pathogen is not uniformly distributed throughout the melon-producing regions of the world, it is very widespread.

In India, Fusarium wilt was first reported from the Maharashtra in 1955 and now it occurs in almost all the states. The disease is reported to be serious in Punjab (Waraitch *et al.*, 1976).

Symptoms

Fusarium wilt of watermelon is caused by the fungus *Fusarium oxysporum* f.sp. *niveum*. The fungus also attacks summer squash but not muskmelon or cucumber. The plant is attacked in all the stages of its growth. Germinating seeds may rot in the soil. The cotyledons may wilt. Plants infected early in their development i.e. very young stage of seedling often damp-off at the soil line. Small leaves of infected young plants loose their green colour, droop and wilt or be stunted in growth. Older plants may first exhibit temporary wilting (flagging down of leaves) only during the hot period of the day but will die within a few days. The wilting normally progresses slowly. Wilt symptoms develop in one or more lateral vines, starting at the tip. If cut back the epidermis and cortical tissue (bark) on a section of the main stem slightly above the soil line, a light brown discoloration of the vascular tissue (area just beneath the epidermis) may be seen. Mature plants show typical wilt symptom with vascular browning, gummosis and tyloses in xylem vessels. In wet weather, a white to pink fungal growth may be visible on the surface of the dead stems. The disease occurs more severely at 20 – 30^{0}C (Sohi and Sharma, 1998).

Fusarium wilt of muskmelon is caused by the fungus *Fusarium oxysporum* f.sp. *melonis.* The disease is rated as one of the serious disease that capable of causing up to 90 per cent or more mortality (Sohi and Sharma, 1998). The fungus infects only muskmelon, crenshaw melon, and honeydew melon. Although plants may be affected in any stage of development, but the disease is most common on mature plants. On young seedlings, a hypocotyl rot and damping-off may occur. In older plants, there is marginal yellowing progressing to a general yellowing of the older leaves, and wilting of one or more runners. In some cases, sudden collapse occurs without any yellowing of the foliage. On stems near the crown of the plant, a linear, necrotic streaks or lesion may develop externally, extending up the plant and usually on one side of the vine. Streaks are at first light brown, turning yellowish tan, then dark brown with age. This symptom is diagnostic for the disease. One runner on a plant may wilt and collapse, with the rest of the runners remaining healthy. A gummy, reddish exudate may ooze from these lesions, but this may also be caused by gummy stem blight and insect injury. Vascular discoloration should be evident and is very diagnostic. Mature plants often wilt severely (collapse) late in the season because of the fruit load stress. A white to pink fungal growth may develop on infected stems during wet weather. Palodhi and Sen (1979) studied the role of tyloses in wilting. Optimum growth of the fungi was recorded at 26^{0}C and maximum symptom expression occurs at 18-22^{0}C (Sohi and Sharma, 1998).

Fusarium wilt collapse should not be confused with sudden wilt of melon. Sudden wilt is a complex disease associated with plant stress brought on by heavy fruit set, cool evening soil temperatures followed by warm and sunny days, feeder root loss caused by soilborne fungi (may include *Verticillium* and other species), and virus infection (primarily cucumber mosaic and watermelon mosaic, but also papaya ring spot W strain and zucchini yellow mosaic virus).

Fusarium wilt and root rot of cucumber caused by *Fusarium oxysporum* f.sp. *cucumerinum* produces pre- and post- emergence damping-off of seedlings and wilting of mature plants. Young infected plant show dark brown cortical lesions followed by soft decay of collar tissues at the base of the stem. Wilting symptom first develop on single lateral vine followed by total collapse of infected plants. Xylem vessels show necrosis up to the seventh node. Favourable temperature for the pathogen is below 20^{0}C (Sohi and Sharma, 1998).

The causal pathogens and perpetuation

Fusarium oxysporum Schlecht. Snyder & Hansen [with the following forma species (f.sp.) responsible for wilt of different cucurbits]:

Fusarium oxysporum f.sp. *benincasae* Ying C. Wu & S.Z. Wang.

Fusarium oxysporum f.sp. *cucumerinum* Owen.

Fusarium oxysporum f.sp. *lagenariae* Matuo et Yamamota.

Fusarium oxysporum f.sp. *luffae* Kawai *et al.*

Fusarium oxysporum f.sp. *momordicae* Sun & Huang.

Fusarium oxysporum f.sp. *niveum* (E. F. Smith) Snyder & Hansen.

The causal fungus survives from season to season in soil as mycelium as well as chlamydospores (thick-walled modifications of the mycelium), in old infected vines or on seeds. The pathogen is both externally and internally seed borne. Internal infection is limited to the area immediately beneath the seed coat and does not infect the embryo. The fungus can live on dead plant material (saprophytically) or on the roots and stems of other plants, such as tomatoes and several weeds. These serves as the primary inoculum source for the next disease cycle. Infection occurs through the root tip by direct penetration, natural openings or through wounds. Root-feeding larvae of the striped cucumber beetle may increase the incidence of wilt. Eventually, the fungus invades and plugs the plant's water conducting vessels, reducing water movement and leading to plant wilt and death. The pathogen can be spread to other fields through infested on farm equipment, infected crop debris, wind blown soil and irrigation water (Gour *et al.*, 2008). It can also be spread by infected plant material or contaminated seed.

In case of *Fusarium oxysporum* f.sp. *melonis* (*Fom*) four races (called races 0, 1, 2, and 1,2) have been accepted worldwide based on specific resistance genes found in melon differentials that the pathogen overcomes. The only described race of *Fom* in the United States is race 2. In 1985, race 1 was reported from Maryland, and in 1987, race 0 was discovered in Texas. In 1992, race 1 was recovered from collapsed melon fields in New York. But race 2 still remains the most widely distributed race in the United States. In contrast, race 1 is the most common race in Europe and the Middle East. Until recently, race 1,2 had only been reported from France, but is now known to occur in Maryland. Through the early breeding efforts of Henry Munger, two Fusarium wilt resistant varieties (Iroquois in 1944 and Delicious 51 in 1951) with *Fom*-2 resistance gene were released to meet the early needs of New York growers.

Epidemiology

Environmental and soil conditions are important for infection and in symptom expression. Disease incidence and severity increase during warm and dry weather conditions. Warm soil temperatures favour the disease development (Gour *et al.*, 2008). Seedling injury is severe at 20 to 30°C temperature. Wilt development is also favoured by a temperature of about 27°C. No infection occurs at temperature below 15°C and above 35°C. At high soil temperatures, plants become infected but may not wilt; rather they develop severe stunting. Low soil moisture is congenial for the pathogen and accentuates the wilting symptom. High nitrogen, especially NH_4 - nitrogen, and light, sandy, slightly acidic soil (pH 5-5.5) favour disease development. Ramasamy and Prasad (1975) stated that higher doses of nitrogen increase the disease intensity whereas higher doses of potassium reduce it. Kesavan and Prasad (1974) recorded the presence of more cucurbitaceae in resistant varieties of muskmelon than in susceptible ones, but direct correlation has not been established.

Disease management

a) Resistant varieties, along with rotations with non-hosts, offer the most reliable control of this disease. Resistant varieties should be planted in the same field only once every 5 to 7 years and susceptible varieties no more than once every 15 years. Radhakrishnan and Sen (1985) reported Durgapura Madhu and Punjab Sunehri as resistant varieties of muskmelon.

b) Use of disease free seed and destruction of diseased debris can keep the disease under check.

c) Liming the soil to increase pH about 6.5-7.0 decreases wilt severity.

d) Addition of cellulosic materials (straw) to soil reduces survival of the pathogen in soil (Kannaiyan and Prasad, 1975).

e) Seed borne inoculum can be reduced by hot water treatment at 55°C for 15 minutes (Sohi and Sharma, 1998). By treating the seed with Benomyl or Carbendazim 50% WP at 1-1.5 g/kg seeds or *Trichoderma harzianum* @ 6g/kg seeds the disease may be kept under check. *Trichoderma viride* @ 4g/kg seeds helps in reducing the seed borne inoculum (Gour *et al.*, 2008).

f) The disease can be checked to some extent by drenching the soil with Captan 50% WP (0.2 to 0.3% solution). This should be repeated twice or thrice.

g) *Trichoderma viride* formulation 2 kg mixed with 50 kg of farm yard manure per acre is effective (Gour *et al.*, 2008).

2.5.1.6 Alternaria leaf blight

Alternaria leaf blight is found primarily on muskmelon, but may occur on watermelon, cucumber, gourds, pumpkin and squash (Seebold, 2010, Gour *et al.*, 2008). This disease affects foliage and sometimes fruit. Alternaria leaf blight does not commonly infect fruit but can reduce yield and quality through reduced plant vigor and sunscald of exposed fruit. The disease was recorded on muskmelon, watermelon, bottle gourd, snake gourd, cucumber and vegetable marrow in Punjab and Rajasthan (Khandelwal and Prasada, 1979, Chahal *et al.*, 1970). Bhargava and Singh (1985) recorded a wide range of disease intensity from Rajasthan on various cucurbits. The disease intensity varied from 30.2 to 90.7 per cent on water melon, 30.2 to 40.8 per cent on bottle gourd, 12.5 to 60.3 per cent on pumpkin, 12 to 22.2 per cent on ridge gourd, and 15.3 to 29.0 per cent on bitter gourd. They also reported 77.7 per cent yield losses on bitter gourd, 65.3 per cent on bottle gourd, 80 per cent on pumpkin, 69 per cent on ridge gourd, and 88.3 per cent on water melon.

Symptoms

Disease symptoms initially appear on mature leaves near the crown of the plant as small necrotic spots, which rapidly increase in number and size. The spots may be surrounded by a yellow halo. These lesions expand to form large, irregular brown spots with a concentric ring or target board pattern. Expanding lesions may merge to form large, blighted areas. The blighting symptom or burning effect is prominent especially on the water melon. As symptoms progress, leaves curl and die, leading eventually to plant decline. Fruits are not commonly infected but can suffer from sunscald due to leaf loss. Gour *et al.* (2008) reported that fruit infection on musk melons causes circular, brown and sunken spots resulting in dry rot. The fruit rot phase of the disease needs more study.

The causal pathogen and perpetuation

Leaf blight is caused by *Alternaria cucumerina* (Ell. & Ev.) Elliot. This fungus perpetuates as a saprophyte on decaying crop debris in soil or in weeds and other crops (Gour *et al.*, 2008). The disease dissemination occurs from plant to plant through conidia that are carried by wind and splashing water from diseased plants to disease free plants (Seebold, 2010). The long distance dissemination of the disease occurs through wind currents whereas the disease can be spread within the field by splashing water. The role of collateral hosts and seeds as carriers of the pathogen needs to be investigated appropriately (Sohi and Sharma, 1998).

Epidemiology

Germinating spores can penetrate the host directly, as well as through wounds and natural openings. Wet rainy weather favour the disease development. The disease is more severe in wet and warm temperatures condition (Gour *et al.*, 2008, Seebold, 2010). A temperature of 25 to 30°C and relative humidity of 92 to 100 per cent are optimum for growth and sporulation of the fungus (Khandelwal and Prasada, 1979). Chopra and Jhooty (1974) studied biochemical changes in resistant and susceptible varieties of water melon.

Disease management

a) Crop rotation with non-host crop for at least 2 to 3 years is helpful for reducing the disease incidence (Gour *et al.*, 2008, Seebold, 2010, Kennelly, 2012).

b) Good sanitation practices, such as cleaning up crop debris at the end of the growing season, proper drainage during growing season are important.

c) Use of drip irrigation instead of overhead sprinklers is helpful to manage the disease.

d) Muskmelon cultivars that have some level of resistance should be cultivated; newer varieties are believed to be somewhat more resistant to Alternaria leaf blight than older, traditional cultivars

(Seebold, 2010). Honeydew melons are highly resistant to Alternaria leaf blight (Kenndelly, 2012).

e) Borax wash (2.5 per cent) at 45°C for 30 second or at 40°C for 2 minutes before packing of the fruits prevents fruit rot. Captan, Maneb and Ziram are also effective at low temperatures (Sohi and Sharma, 1998).

f) Fungicides can be used to control the disease effectively. Preventive sprays are efficacious but are only essential in fields with a previous history of the disease. Three sprays of Carbendazim 50% WP or Benomyl 50% WP @ 0.2% at 15 days intervals for managing the disease of watermelon has been recommended under Bangalore conditions (Rai *et al.*, 2008). The disease can be managed effectively by spraying Mancozeb 75% WP @ 0.25% or Copper oxychloride 50% WP @ 0.3% (Gour *et al.*, 2008).

2.5.1.7 Pythium fruit rot

Fruit rot is a fungal disease and is incited by *Pythium* spp., which is very common in India. The disease is also known as blossom end rot (Sohi *et al.*, 1976), cottony leak (Singh and Chohan, 1977) and fruit rot. It occurs in almost every locality and field during the rainy season (July to August). In Punjab, 60 per cent or even more fruits of muskmelon and water melon occasionally rot after rains because of cottony leak (Singh and Chohan, 1977). In Karnataka, the disease affects nearly all of the cucurbits; however it is a major problem in bottle gourd. The disease was reported on sponge gourd, snake gourd, pointed gourd, kheera, bottle gourd and bitter gourd. Fruits may rot during transit and storage also.

Symptoms

The disease appears as luxuriant cottony mycelial growth on the affected fruits which looks like wrapped in cotton. The fruits in contact with soil suffer most. The first noticeable symptom observed on fruits surface as small, water soaked lesions, which gradually turn dark green and develop into a watery soft rot. Under humid atmospheric condition, the diseased tissues are covered by a profuse, whitish, fluffy mycelial growth. In certain cases, the whole fruit surface may be covered by mycelial growth. In watermelons, the decay often starts at the blossom end and progress towards stem and cover the entire tip region. On the margin of the cottony growth the skin of fruit looks dark green and water soaked. The area, which is without any aerial growth of the fungus, indicates the 'killing in advance' activity of the pathogen. The tissues in the interior of the fruit become watery, turns brownish and soft, and the decaying matter emits a bad odour. The disease is common in the field where the fruit lying on the ground or hangs near the soil level. The disease spreads among the fruits during transit and storage. On the fruits, infection generally occurs when there is some injury to the skin, by soil particles, excess wetness around fruits, or insect bites or injuries may be happen during intercultural operation. Chattopadhyay and Sengupta (1952) and Saha *et al.* (2002) recorded detail symptoms of the disease on pointed gourd.

The causal pathogen and perpetuation

The disease caused by the fungal pathogen *Pythium* spp. *Pythium butleri* Subramaniam and *P. aphanidermatum* (Edson) Fitzp. are commonly associated with the fruit rot in Karnataka (Singh and Chohan, 1977, Sohi *et al.*, 1976), whereas *P. debaryanum* Hesse, *P. ultimum* Trow and many other fungi also cause the disease either singly or in association.

The mycelium of the *P. aphanidermatum* consists of intracellular, much branched hyphae. The hyphae measure 2.8-7.5µm, mostly 4.0-6.0µm in diameter, produces much lobbed, branched sporangia, which may be less than 50-100µm in length. Oogonia are spherical smooth walled, terminal on lateral hyphae, 19-29µm in diameter. Oospores are aplerotic, single, with moderately thick wall and 17-19µm in diameter (Gupta *et al.*, 2001).

The pathogen is soil-borne and can be perpetuated in wild cucurbits crops in absence of hosts.

Epidemiology

The pathogen grows at a temperature range of 15 to 40°C, with optimum growth at 30 to 35°C and 100 per cent relative humidity (Sohi and Sharma, 1998).

Disease management

a) Though some of the cucurbit fruits consumed directly, it is not always advisable to recommend application of fungicides to prevent infection of fruits.

b) Soil disinfection will be uneconomical.

c) By changing the method of cucurbit cultivation like use of mulching or scaffolding can keep the fruits detached from contact with the soil that can reduce the chance of infection.

d) Hygienic transportation and storage may be used to avoid this fruit rot.

2.5.1.8 Fruit and Vine rot of Pointed gourd

Pointed gourd (*Trichosanthes dioica* Roxb.) is an important and highly accepted cucurbitaceous vegetable in West Bengal. As the crop prefers warm condition and major part of the growth phase passes through rainy season it suffers from large number of diseases. In West Bengal fruit and vine rot of pointed gourd appears every year and causes severe damage of the crop. Destruction of the entire field is also common in the rainy season (Saha and Khatua, 2004, Guharoy *et al.*, 2006).

Symptoms

In West Bengal the pointed gourd crop is grown in soil bed or raised on scaffold (*macha*). Sometimes the crop is maintained as ratoon crop. Symptom appears early in ratoon crop than new or fresh crop. After the vine growth covered the field a few vines dried in the field following a shower. Oozing and drying of gummy substance found in one or two internode(s) of such dried vines. Drying occur above the point of oozing. Affected vine tissue turned brown. If there is rain or cloudy weather white mycelial growth of the fungus could be seen over the brown region. In April and May, fruit infection started. Infection starts at any point of the fruit usually near the middle portion. In moist condition affected area appeared water soaked and covered with white fungal growth. After the monsoon rain starts in June-July, the field soil remained moist. Frequent rains and cloudy conditions favoured development of disease. More and more vines and fruits were infected. Intensity of fruit infection became more. During the period of continuous rain, infection initiates on leaves near the point of attachment of leaf lamina with petiole, causing leaf blight symptom. Ultimately the entire crop may be damaged. Crop grown on *macha* or scaffold suffered less. Fruit rot followed by stem (vine) rot becomes severe only in rainy condition (Saha and Khatua, 2004, Mondal *et al.*, 2013).

The causal pathogen and perpetuation

Under humid condition in field, the pathogen produces abundant sporangia on the infected tissues. If small bits of artificially inoculated and colonized fruit tissue, naturally infected fruit or stem tissue are put in a petri dish in half submerged condition in water, the fungus produces profuse mycelia and sporangia. Sporangiophores are indeterminate undifferentiated, sympodially branched. The sporangia are ovoid to elliptical measuring 43-140μm × 21-74 μm (average 85.8μm × 47.6μm). After 48 to 72 hours of incubation of the infected tissue in water, zoospores are released in water through the opening at the top of the sporangia.

The interesting phenomenon is that the sporangia proliferate 1-3 times at same point. Growth of sporangiophore may continue through the empty proliferated sporangia to produce new sporangia. Empty sporangia remain attached with the sporangiophore. Sporangia also germinate directly by producing germ tube (1-2) in humid condition but in absence of free water. The fungus produce chlamydospores which are usually intercalary formed singly or in chain. Occasionally chlamydospores are produced on the terminal end of the hyphae. They are elliptical in shape, extremely variable in size, measuring 2.4-18.8 μm × 2.4-13.7 μm. Oospore is not formed in affected tissue or in culture medium (Mondal *et al.*, 2013). Earlier the causal fungus was known as *Phytophthora cinnamomi* Rands (Khatua *et al.*, 1981). Letteron the fungus has been identified as *Phytophthora melonis* Katsura based on morphological characters (Waterhouse, 1970) and the Restriction Fragment Length Polymorphism (RFLP) sequencing of ITS region (Guharoy *et al.*, 2006).

Through modification of medium originally proposed by Tsao and Ocana (1969) for *Phytophthora* sp., a suitable medium was developed by Mondal *et al.* (2013) for isolation of *P. melonis*. The composition of the medium is oat meal 17 g, vancomycin 200 mg, natamycin 10 mg, carbendazim 25 mg, agar agar 20g, distilled water 1000 ml. This medium is very much suitable and easy for isolation of *P. melonis* in pure form (Mondal *et al.*, 2013).

Epidemiology

Disease appears late (April-May) in newly planted crop than ratoon crop (March). Crop grown on soil bed suffers more than the crop raised on scaffold. Cloudy weather, rainy days and water stagnation favoured rapid spread of the disease. During the period of frequent rains or rainy and cloudy weather condition along with prevailing warm or high temperature for a few consecutive days (6 days or more), the progress of the disease become very fast. Such conditions favour rapid and abundant production of sporangia. During rainy days, free water may be present in and around the diseased tissue, particularly in case of crop raised on soil bed. Such condition helps in production and release of zoospores. The zoospores are distributed over the field by rains. During heavy rain or continuous rain the water stands in the field filling the irrigation channel for a period of one to few hours. If there is incidence of fruit and vine rot disease in the field and the atmospheric temperature above 28°C, the disease appears in severe form after the rain. During the rainy day, if the water from infected field flows to another field helps in distribution of inoculum, particularly the zoospores produced in presence of free water. Continuous cultivation of the crop i.e. pointed gourd followed pointed gourd in a same piece of land increases disease severity. Fruit and vine rot disease appears late in the season in isolated field may be due to lack of inoculum source. Diseased vines and fruits in the field and irrigation channel increase disease severity. Male plants are less susceptible than the female plants (Mondal *et al.*, 2013).

Disease management

a) Ratooning of the crop should be avoided (Mondal *et al.*, 2013).

b) Crop should be raised in scaffold (Mondal *et al.*, 2013).

c) Crop rotation with non-host crop may delay incidence and reduce severity of the disease (Mondal *et al.*, 2013).

d) Proper sanitation measures should be taken to avoid the disease. Infected fruits should not be thrown in irrigation channel. Crop should be grown in well drained field. Collection and destruction by burning of infected fruits and vines is important to manage the disease (Bhattacharya *et al.*, 2006, Mondal *et al.*, 2013).

e) Spraying of Carbendazim and a combination product of Metalaxyl + Mancozeb promote vine growth and production of small, soft, tender vine branches above the crop canopy. Repeated spraying of such fungicides increases disease severity, in rainy days, instead of disease reduction. Carbendazim is not effective against *P. melonis*, but farmers are attracted by the plant growth stimulative action of the fungicide (Mondal *et al.*, 2013). Spraying of Metalaxyl 8% + Mancozeb 64% WP, Azoxystrobin 25% SC, Copper oxychloride 50% WP, Fenamidone 10% + Mancozeb 50% WG, Fosetyle-Al 80% WP at the recommended doseges at 10-12 days intervals recorded efficacious (Annonymons, 2010). Alternate use of Copper oxychloride 50% WP @ 0.4%, Metalaxyl 8% + Mancozeb 64% WP @ 0.25%, Cymoxanil 8% + Mancozeb 64% @ 0.25% at 15 days interval is effective (Bhattacharya *et al.*, 2006). Some antibiotics (Streptomycin sulphate + Tetracycline hydrochloride, Streptomycin sulphate, Oxytetracycline hydrochloride, Ofloxacin and Norfloxacin) recorded effective to inhibit sporangia formation at 100ppm in laboratory condition by Khatua *et al.* (2013). Adverse effect of Chloroquin phosphate and Copper sulphate were also recorded by them.

2.5.1.9 Net blight or Leaf blight

This is an important disease of pointed gourd. The disease also locally known as *suji dana haja* (*suji* is a granulated but not pulverized product of Indian wheat, *dana* means seed, *haja* is a type of skin disease) due to the presence of small white sclerotial growth on leaves and stem. Sometimes farmers called it *chak poka* (*chak* means 'beehive like shape', *poka* means 'insect') because of appearance of symptom as beetle feeding on leaves (Khatua and Saha, 2004, Mondal *et al.*, 2014). The disease was recoreded from Nadia and North 24 Parganas of West Bengal (Mondal *et al.*, 2012).

Symptoms

The disease starts as water soaked angular spot, delimited by veinlets on lower side of the leaves. At that time, no symptom appears at the upper surface of the leaf. In time, such leaves dried up. During rains, rotting occurs in infected zone of the leaves and rotted tissues are washed out by rain leaving the network of veins. The first noticeable symptom of the disease in the field is drying of older leaves in patches.

During rainy days or sunshine day following rains at night, white small sclerotial growth is observed on the diseased tissue and nearby healthy tissue early in the morning. Thin hyphal filaments are also found on the diseased tissue. After sunrise, the hyphal and sclerotial growth apparently disappears. In advanced stages of the disease all the leaves dry up and axillary buds are destroyed leaving the stem bare (Saha, 2002).

The disease appears from August onwards particularly during the rainy days where crop is grown in soil beds.

The causal pathogen

The disease cauing agent is a small sclerotia forming fungus *Rhizoctonia solani* Kuhn (Saha, 2002) causes leaf blight of radish in the same season (Khatua and Maiti, 1982). The pathogen also causes leaf blight of bottle gourd as reported from Nadia district of West Bengal by Mondal *et al.* (2012).

Disease management

a) Spraying of mixture of Mancozeb 75% WP (0.2%) and Carbendazim 50% WP (0.1%) once or twice is effective in controlling the disease (Khatua and Saha, 2004).

b) Crop rotation with non-host crops is recommended.

2.5.1.10 Fusarium root rot

Fusarium root rot of squash, also known as foot rot or crown rot attacks almost all the cucurbits. Root rot of squash was first described from South Africa in 1932. In 1939, widespread reports were received from New York, Connecticut, and Massachusetts regarding a serious disease of summer squash, which was subsequently traced to a particular lot of seed grown in Oregon. The disease has also been reported in Australia and Canada. While its distribution in the United States is limited, it still remains a concern for cucurbit growers. The crown and foot rot phase of disease has occurred sporadically in New York since 1939, but in 1995 the fruit rot phase was experienced in two widely separated countries.

Symptoms

The above ground symptoms of this disease are similar to those of vascular wilt caused by *Fusarium oxysporum.* The first symptom generally noticed in the field is wilting of the leaves. Within some days, the entire plant may wilt and die. If the soil is removed around the base of the plant, a distinct necrotic rot of the crown and upper portion of the taproot is evident. The rot develops first as a light-colored, water-soaked area that becomes gradually darker. It starts in the cortex of the root, causes cortex tissue to slough off, and eventually destroys all of the tissues except the fibrous vascular strands. Infected plants break off easily about 2-4 cm below the soil line. The fungus normally is limited to the crown area of the plant. The main and lower portions of the taproot are not affected, except under extremely wet conditions. Similarly, the stem is not affected, except for the lower 2-4 cm immediately above the soil line.

The above ground symptoms of this disease may be confused with vascular wilt caused by *Fusarium oxysporum.* But there is sudden wilting of the plant in mild season with dark brown cortical soft decay at the base of the stem. The below ground parts of the infected plants are found disintegrated. Frequently there is invasion of insects in such roots, and secondary infection by other organism likely fungi may also be found. Profuse sporulation occurs on infected tissues during humid weather condition (Sohi and Sharma, 1998). Plants showing symptoms develop numerous sporodochia and macroconidia (spores) giving the mycelia a white to pink colour on the stem near the ground surface. Fruits are attacked at the fruit-soil interface; the severity of the fruit rot depends on soil moisture and the stage of rind maturity at the time of infection.

The causal pathogen and perpetuation

Fusarium crown and foot rot is caused by *Fusarium solani* f.sp. *cucurbitae* Snyder & Hansen. Two races of the fungus have been described.

Race 1 is distributed worldwide and causes root rot, stem rot, and fruit rot. It is responsible for the disease in New York. Race 2 causes only fruit rot and has been reported only in California and Ohio. Recent work, however, has separated these two races into different mating populations which most probably mean that the two races represent different species. *F.solani* f.sp. *cucurbitae* forms all three asexual spore types (microconidia, macroconidia, and chlamydospores) typical of *Fusarium* by which the fungus can overwinter. The fungus also produced sexual stage i.e. *Nectria haematococcus* Berk & Br. (=*Hypomyces solani* Reinke and Berth.), (Sohi and Sharma, 1998).

The fungus is both internally or externally seed borne. However it actually survives for only one to two years in seed. Infection does not appear to affect seed viability or germination. Although the fungus produces abundant chlamydospores, it apparently survives for only two to three years in soil, which is much less than *F. oxysporum* or even other formae speciales of *F. solani*.

Epidemiology

Plants and fruit of any age can be infected. Muskmelon, squash melon and bitter gourd are most severely affected by this disease than pumpkin. The pathogen exhibits host specificity for all cucurbits. Larger pumpkins are more susceptible than smaller sized varieties. The smaller sized varieties matured sooner in the field, and the time of infection and environmental conditions may not have been conducive for fruit infection to occur. The disease severity primarily depends on soil moisture and inoculum density.

Disease management

a) Removal of diseased plants from the field is important.

b) Use of healthy and disease free seeds can avoid disease situation in main field.

c) Hot water treatment of seed at 55^0C for 15 minutes (Roychaudhury and Verma, 2000) or seed treatment with systemic fungicides like Carbendazim 50% WP @ 2g/kg seed reported efficacious.

d) A three to four year crop rotation with non-host crop is adequate for managing the disease (Roychaudhury and Verma, 2000).

e) Spraying of the crop with Captan 50% WP (0.25%) is effective.

f) Cultivation of resistant varieties is recommended.

2.5.1.11 Alternaria Fruit rot of Pointed gourd

Symptoms

The disease is prevalent in diara area. Black uniformly spreading lesions appear on pointed gourd fruit. The fungus gradually forms a cavity in the infection court, which later appears as sunken lesions. Finally it produces deep brown spore mass. Later the infection extends deep inside the flesh. The fruit become pale yellow leaving a deep brownish-black target board like spot on the skin and dries up due to loss of turgidity. Brownish mycelia are observed inside the rotten fruits. Whole fruit including seeds are found to rot emitting unpleasant odour (Khatua and Saha, 2004). The pathogen also caused severe leaf spot disease on cucumber. In this case, lesions ranged in size of a pin point to over 5 cm in diameter, with necrotic tissue on most of their area and a surrounding yellow zone (Vakalounakis and Malathrakis, 1988).

The causal pathogen and pepetuation

The disease caused by the fungal pathogen *Alternaria alternata* (Fr.) Kiessler. Conidia of the fungus are obclavate, dark, muriform and measured 30-75×15-25 μm in size. This pathogen hardly ever attack young, vigorously growing plants. They live from one season to the next in infected crop debris and wild cucurbits (Sahu Kritagyan and Singh, 1980).

Epidemiology

The pathogen causing leaf spot on cucumber grew satisfactorily on PDA at temperatures between 5°C–40°C and spore germination occurred in the range less than 10°C to over 37°C. Optimum temperature in both cases is near 26°C (Vakalounakis and Malathrakis, 1988).

Disease management

There are some chemical and non-chemical control measures, which may help to reduce the disease incidence.

a) A single postharvest water dip of 20 min. at 48 °C will reduce the incidence of *Alternaria* fruit spot (Nishijima, 1993).

b) Chlorothalonil 75% WP or Mancozeb 75% WP @ 0.2% sprayed once every two weeks can reduce *Alternaria* fruit spot by about 50%. However, an orchard spray program alone does not provide adequate economic control necessary for surface shipment to export markets. The water dip will be helpful for reducing the incidence (Nishijima, 1993).

c) In Green house condition, Iprodione 50% WP @ 0.1%, Prochloraz-manganese-complex, Chlorothalonil 75% WP @ 0.2%, Dichlofluanid, Guazatine, Maneb and Etem were reported to be effective for controlling the disease caused by *A. alternata* (Vakalounakis and Malathrakis, 1988).

2.5.1.12 Rhizoctonia fruit rot

The disease was recorded by Som and Bandyopadhyay (1981) during kharif season of 1979 in a pointed gourd field of Bangaon sub-division, West Bengal where jute was cultivated previously. The loss was estimated to be 20-25% (Khatua and Saha, 2004).

Symptoms

The disease appears as yellowish orange discolouration on the skin of green fruit, which later changes to light brown water-soaked spots spread on the fruit surface that become necrotic. These spots enlarge and gradually cover whole fruit, which ultimately rots and shrivels with the loss of moisture. The affected tissue becomes hard and dotted with sclerotia of the pathogen. The pulp of the fruits becomes brown to black with mycelial growth with sclerotia whereas no pycnidia are noticed on fruit pulp. Leaves touching the soil can be infected. The fungus will web to adjacent leaves hastily under high humid conditions and will rot such leaves rapidly.

The causal pathogen

Som and Bandyopadhyay (1981) identified the causal pathogen as *Rhizoctonia bataticola* (Taub.) Butler.

The fungus persists in the soil as hyphae and sclerotia. Sexual stage of the pathogen is very rare. The young hyphae are colorless but turn brown with age. Soil particles frequently adhere to the infected part of the plant when removed from the soil because of the coarse, brown mycelium. Under microscopic observation, hyphal branches are at 90 degree angles to the parent hypha and there is a cross-wall and a constriction of the cell at the base of each branch.

Epidemiology

The fungus favours warm soil temperatures 12 to 32 °C for growth and infection. Moderate to high soil moisture favors growth of the fungus and development of the disease.

Management

a) No effective biological control strategies have been developed for Rhizoctonia fruit rot.

b) By staking plants and using plastic mulches barrier between fruit and soil the incidence of the disease can be minimized. Varieties highly susceptible to Rhizoctonia fruit rot should be avoided, especially if plastic mulch is not used.

c) Fungicides like Oxycarboxin 20% EC @ 0.15% or Validamycin 3% L @ 0.2% reduce fruit rot losses of cucurbits.

2.5.1.13 Stem rot / Collar rot

This is a minor disease of cucurbitaceous vegetables. The important cucurbits like watermelon, winter squash, cantaloupe, cucumber, bottle gourd, muskmelon, pumpkin, squash etc are reported to be infected by the pathogen causing stem rot. The disease is most common in the tropics, sub-tropics and other worm temperate regions of the world.

Symptoms

The pathogen infects any portion of the vine in the field. The first sign of the disease is a mid-day wilting of the plant. The leaves turn yellow and within a few days, the plant completely wilts and dies due to girdling of the stem at the soil surface (Gour *et al.*, 2008). It causes rotting of the stem portion that is in close contact with soil. The affected tissue becomes covered with white mycelial growth. Sclerotia are formed on the infected tissue and on nearby soil. Severely infected vine may dry up. Initially dark brown lesions formed on the stem at or just beneath the soil level, and are undetectable. The first visible symptoms are progressive yollowing and wilting of leaves/vines following production of white, fluffy, fan-shaped mycelium on infected tissue and in the soil. The pathogen may infect fruits, leaves and flowers under favourable environmental conditions.

Seedlings are very susceptible and die quickly after infection. The pathogenic infection develops gradually in older plants due to formation of woody tissue, and infected plants die eventually. Invaded tissues are pale brown and soft, but not watery.

Maximum damage occurs when the fungus attacks the roots of root knot infected plants, causes root rot resulting into death of the entire plant (Khatua and Saha, 2004).

The causal organism and perpetuation

Singh and Seth (1974) recorded that the fungus, *Sclerotium rolfsii* Sacc. (Teleomorph / sexual stage: *Athelia rolfsii* (Curzi) Tu & Kimbrough, syn. of sexual stage: *Corticium rolfsii, Pellicularia rolfsii*) causing collar rot disease.

Rapidly growing, silky-white hyphae of *S. rolfsii* tend to aggregate into rhizomorphic cords. A white, fluffy, fan-shaped, acutely branched, septate mycelial expanse may be observed growing outward. Sclerotia (0.5-2.0mm diameter) begin to develop after 4-7 days of mycelial growth. Initially sclerotia are whitish in colour, which quickly melanize to a dark brown colouration resemble mustard seeds having spherical to irregular in shape. Sclerotia, which form on hosts, are about smooth textured, whereas those produced in culture may be pitted or folded. Sclerotia of relative uniform size are produced on the mycelium. Sclerotia are the resting structure and contain viable hyphae that serve as primary source of inoculum for disease development. The mycelium servives best in sandy soil, while the sclerotia servive best in moist and aerobic conditions found at the soil surface (Aycock, 1966). The pathogen survives in soil for many years and the disease spreads by different means i.e. movement of infested soil or by surface water (Gour *et al.*, 2008).

The sexual stage of the fungus (*Athelia rolfsii*) occastionally produces basidiospores at the margin of lesions under humid conditions, but this form is not common. Reproduction and spread of the organism by this spore under field conditions is unknown (Sarma and Singh, 2002).

Epidemiology

Sclerotia serve as the principle overwintering structures and primary inoculum for the disease. Persisting near the soil surface, sclerotia may exist free in the soil or in association with plant debris. Those buried deep in the soil may survive for a year or less, whereas those at the surface remain viable and may germinate in response to alcohols and other volatiles released from decomposing plant material. This soil borne, polyphagous fungal plant pathogen has wide host range that serve as collateral hosts (Aycock, 1966).

High temperatures and moist conditions are associated with germination of sclerotia and development of the disease (Gour *et al.*, 2008). Maximum mycelial growth occurs between 25 and 35 °C. High soil moisture, dense planting, and frequent irrigation promote infection.

Since *S. rolfsii* (asexual stage) does not produce spores, dissemination depends on movement of infected soil and infected plant material. Use of contaminated equipment and machinery may spread sclerotia to uninfested fields.

Disease management

a) Deep plowing (at least 20cm) serves as a cultural control tactic by burying sclerotia deep in the soil.

b) Removal and burning of infected plants can prevent the fungal infection. Crop rotation also helpful (Gour *et al.*, 2008). Due to the polyphagous nature of the pathogen selection of non-host crop is very much important to manage the fungus through crop rotation.

c) Soil solarization is effective to manage the disease by reducing inoculum. Covering soil with transparent polyethylene sheets during the hot season increases soil temperatures and kills sclerotia. Soil solarization in combination with the application of *Trichoderma harzianum* is more effective to manage the disease.

d) There are some reports about the use of black plastic mulch in controlling the disease.

e) Seed treatment with *Trichoderma viride* @ 4 g per kg seed will help in reducing the disease. Soil treatment with *T. viride* recorded effective too (Kulkarni and Kulkarni, 1994, Virupaksha Prabhu *et al.*, 1997, Singh *et al.*, 2013), while soil application of *Azotobacter chroococcum* and *Pseudomonas fluorescens* during transplanting/sowing and 25 days after transplanting @ 10g/plant recorded effective (Mahato *et al.*, 2014). Seed treatment with *Trichoderma viride* @ 4 g per kg seed followed by soil application of 2kg *T. viride* mixed with 50 kg farm yard manure per acre is effective in managing the disease (Gour *et al.*, 2008).

f) Mahato *et al.* (2014) observed effectiveness of vermicompost @ 5t/ha for managing the pathogen causing collar/stem rot disease.

g) Application of plaster of paris on the soil after removal of the infected vines can check the disease. Application of plaster of paris @ 2.5g/infected area of plant reported efficacious (Mondal and Khatua, 2013, Mahato *et al.*, 2014).

h) Mondal and Khatua (2013) recorded some fungicides having inhibitory effect on sclerotial germination of the fungus viz. Carboxin 37.5% + Thiram 37.5% WP, Chlorothalonil 75% WP and Metalaxyl 8% + Mancozeb 64% WP @ 0.2%, 0.2% and 0.25%, respectively. Mahato *et al.* (2014) reported some fungicides (Carboxin 37.5% + Thiram 37.5% @ 0.2%, Metalaxyl 8% + Mancozeb 64% WP @ 0.25%,

Chlorothalonil 75% WP @ 0.2%, Mancozeb 75% WP @ 0.25% and Cymoxanil 8% + Mancozeb 64% @ 0.25%), plant oil (Karanja oil @ 5% v/v) and plant leaf extract (*Murraya exotica* 10% v/v) having good *in vitro* sensitivity against the pathogen.

2.5.1.14 Rhizoctonia root rot

The Rhizoctonia root rot disease was reported from Punjab by Jhooty and Grover (1971). The fungus was pathogenic to all the 10 cucurbits tested by them, causing both pre and post emergence rots. Older plants were less susceptible than younger.

Collar rot of musk melon has been reported by Suhag and Duhan (1980) from Haryana. It causes as much as 50% mortality of the musk melon. The infection usually appears in the field when the plants attain 40 to 50 days of age and the disease gradually starts to spread.

Symptoms

The disease causes light green to yellow green, minute and sunken lesions at the collar region. The depressed infected area extends upwards and may girdle the vine within 15 days, leading to its collapse.

The causal pathogen

The disease is caused by *Rhizoctonia bataticola* (Taub.) Butler.

Epidemiology

The disease caused by *Rhizoctonia bataticola* spreads rapidly from April to June, when day temperatures range between 30 and 40^{0}C (Sohi and Sharma, 1998).

Disease management

Detailed studies are required to develop a suitable management practices. However, some of the measures may be taken as follows:

a) Seed treatment with Carboxin 37.5% + Thiram 37.5% WP @ 0.25% may be helpful for reducing the disease incidence.

b) Crop rotation with non-host crops is recommended.

2.5.1.15 Charcoal rot of fruits

This is a very serious fungal disease of cucurbit fruits. The disease occurs all over India wherever cucurbits are grown.

Symptoms

The fungal pathogen attacks roots, stems and also fruits touching the soil. Water soaked, slightly sunken, brown to rose coloured lesion appear initially on fruit. In advancement of the disease, the entire fruits are affected. The surface of the fruit becomes dark brown or black encrusted with sclerotia. The flesh becomes black and densely impregnated with sclerotia. In advanced stages, the flesh turns characteristic charcoal. Infected seeds turn ash-grey and encrusted with sclerotia. On seedlings black coloured sunken cankers in a concentric ring pattern may appear on the hypocotyls at the time of emergence. The developing canker stunts the plant and wilt. On older plants the crown leaves and runners may turn yellow after infection and eventually die. Characteristically, a water soaked lesion will girdle the vine at the soil level and extend several centimeters upto the stem. Amber coloured droplets may form in the affected part. Within a few days the lesion dries up and turns into light tan colour (Gour *et al.,* 2008).

The causal pathogen and perpetuation

Jhooty and Singh (1971) identified *Macrophomina phaseolina* (Tassi) Goid., [=*Macrophomina phaseoli* (Maubl.) Ashby] causing dark brown to black lesions on muskmelon. Rao (1964 and 1965) recorded same pathogen as a causal agent of charcoal rot of ridge gourd and water melon.

The fungus survives in the soil and overwinters in infected weeds or plant debris (Gour *et al.,* 2008).

Epidemiology

The disease is favoured by high temperatures in moderately wet soil. Moisture and salt stress can predispose plants for infection (Gour *et al.,* 2008).

Disease management

a) Removal and destruction of the affected fruits from the field is helpful for reducing the disease incidence.

b) The incidence of the disease can be reduced by avoiding moisture stress and maintaining plants in good nutritional conditions (Gour *et al.,* 2008).

c) Seed treatment with Carbendazim 50% WP (0.2%) or Carboxin 75% WP (0.2 %) before sowing can check seed-borne inoculums.

d) Application of 2 kg of *Trichoderma viride* by mixing with 50 kg of farm yard manure per acre is efficacious (Gour *et al.,* 2008).

2.5.1.16 Cercospora leaf spots

The disease is widespread all over the tropical and sub-tropical regions of India as well as the world. Although all most all the cucurbit crops are found to be infected with the disease in India, the disease is more common on watermelon, muskmelon, cucumber and *Coccinea indica* (Gour *et al.,* 2008). In favourable weather conditions the disease become more severe. Sometimes the fungus causes devastating damage resulting in heavy loss in fruit yield (Rangaswami and Mahadevan, 2004). In Europe, it is a destructive disease of cucurbits in green house condition.

Symptoms

The disease affects leaves as well as young succulent stems and petioles. Fruits are in general remains unaffected by the disease. The symptom initially appears as minute, water soaked areas on the leaf lamina. These water soaked areas enlarge rapidly to become circular to irregular spots with pale brown, tan or white centers. The margin of the spots becomes purple to almost black in colour. The spots are generally limited by the leaf veins, and they vary in size, from a few mm to about one cm in diameter. A few to over a hundred spots may be seen on a leaf blade, dipending upon the severity of infection. Spots are coalescing, and may form large blotches in advance cases. Severely infected leaves may dry up and ultimately the infected plants die. In snake gourd stem and fruits are also affected. In this case the fungus causes elliptical, greenish, sunken-spots. Several such spots coalesce to cover and girdle the fruit, which rots eventually.

The causal pathogen and perpetuation

The disease is incited by the fungal plant pathogen *Cercospora* spp. Most prevalent species are *Cercospora citrullina* Cooke and *C. lagenariae* Rang. and Chand. (Gour *et al.,* 2008). Other species of *Cercospora* responsible to cause leaf spot disease of cucurbits in India are mentioned below (Rangaswami and Mahadevan, 2004).

Name of cucurbits	*Cercospora* spp.
Trichosanthes anguina L.	*Cercospora citrullina* Cooke, *C. trichosanthes* var. *anguinae* Rang. and Chand.
Momordica charantia L.	*C. momordicae* McRae, *C. citrullina* Cooke
Luffa acutangula (L.) Roxb.	*C. annamalaiensis* Rang. and Chand.
Benincasa hispida (Thumb.) Cong.	*C. citrullina* Cooke
Lagenaria siceraria (Mol.) Standl.	*C. lagenariae* Rang. and Chand.
Citrullus vulgaris Schrad. Ex. Eckl. & Zeyh.	*C. citrullina* Cooke
Cucumis sativus L.	*C. chidambarensis* Rang. and Chand.
Cucurbita maxima Duch.	*C. citrullina* Cooke

The fungal pathogen is known to perpetuate only through the conidial stage. The common source of primary inoculum is infected crop debris. The pathogen may also be perpetuated on perennial weed hosts (Gour *et al.,* 2008).

Fruiting of *C. citrullina* is chiefly epiphyllous. Stomata may or may not be present, if present are small and brown in colour. Conidiophores are 2-30 in a divergent fascicle, pale to very pale brown, paler towards the apex, rather uniform in width or mildly attenuated towards the apex, occasionally swollen at some points, straight to slightly bent or curved, geniculate, multiseptate, simple or occasionally branched, subtruncate at the apex, 20-300 × 4-7 μm in size, conidial scars conspicuously thickened, 3-4 μm wide. Conidia are hyaline, acicular cylindric or slightly cylindro-obclavate or filiform, straight to curved, 1-17 septate, subacute to obtuse at the apex, subtruncate or rounded at the base, and 30-200 × 4-6 μm in size (Hsieh and Goh, 1990, Mukhtar *et al.,* 2013).

There are some reports of leaf spots of cucurbits caused by other fungi. *Helminthosporium rostratum* Drechsler (Krishnamurty *et al.,* 1972) and *Exserohilum rostratum* (Drechsler) Leonard & Suggs (Utikar *et al.,* 1986) causing leaf spots and marginal leaf blight on cucurbits, respectively. Mondal *et al.* (2012) recorded leaf spot (c. o.: *Phyllosticta cucurbitacearum* Sacc.) and leaf blight (c. o.: *Rhizoctonia solani* Kühn) diseases of bottle gourd from Nadia district of West Bengal.

Epidemiology

High humidity along with a temperature range of 26 to 30°C is congenial for reproduction and successful infection of the pathogen (Gour *et al.,* 2008).

Disease management

a) The field and its surrounding should be kept free from perennial weeds to destroy the sources of inoculum.

b) Collection and destruction of the infected crop debris by burning.

c) Proper drainage and aeration facilities in crop field reduce chance of infection and rapid spread of the disease.

d) The seeds should be treated with Captan 50% WP (0.25%) before sowing.

e) Spraying of Captan 50% WP (0.25%), Mancozeb 75% WP (0.25%) or Ziram 27% SL (0.3%) at 10 days interval is effective. Rangaswami and Mahadevan (2004) reported about the efficacy of Zineb 75% WP to check the development of the disease.

2.5.1.17 White mould

The disease was recorded on ridge gourd from Arunachal Pradesh (Bag, 2000), and on pointed gourd from Birbhum district of West Bengal by Khatua *et al.* (2014). The disease on pointed gourd was reported first time from India by Khatua *et al.* (2014).

Symptoms

The first noticeable symptom of the disease in field condition is drying of new vines. Drooping of tips of some vines is also noticed. On careful observation fade green lesions are found on the internodes of those vines. Such vines, if kept in moist condition white mycelial growth come up and spread over the vines within three days. In pointed gourd, infection spread to the leaf also. Prominent white mycelial growth is seen on infected ridge gourd fruits in field condition. Infected pointed gourd fruits become reddish brown and thin mycelial growth cover the infected area. If infected fruits are kept in moist condition prominent white mycelial growth covered the fruits of both pointed gourd and ridge gourd and the rotting of those fruits occurs completely. On the rotted fruits abundant small to large, elliptical, cicular and irregular sclerotia are formed (Khatua *et al.,* 2014).

Pumpkins and some varieties of winter squash are most severely affected by the disease. In this case, the disease does not cause leaf spots but infects both stems and fruit. Water soaked lesion appears on the infected fruits and stems where fluffy white cottony fungal growth is commonly seen. Fruit are often infected through the blossom end of the fruit, especially when the blossom remains attached after pollination, and become rotted and watery. Sclerotia may be found inside these rotted fruit. The disease may occur both in the field and post harvest condition. Stem infections often start where cotyledons are fading or where the plant has been wounded. Small hard and black coloured sclerotia eventually develop that generally embedded in the cottony growth.

The causal pathogen and perpetuation

The causal pathogen, *Sclerotinia sclerotiorum* (Lib.) de Bary produce white mycelium with hyaline, branched and septate hyphae. Black sclerotia near spherical to irregular in shape generally formed within four days of incubation at 25°C. The sclerotia are silvery white in initial stages of development but turned dark later. No host specificity of the fingus was recorded by Khatua *et al.* (2014). Sclerotia of the pathogen produces on infected pumpkins and melons are hard and about a size of a raisin or smaller by which they survive in the soil and in plant debris for 5 or more years. These sclerotia produce small mushroom-like apothecia that release spores under cool wet weather conditions.

Epidemiology

The disease appears mostly in worm/hot, humid/rainy condition. The pathogen has wide host range. The pathogen infects French bean (*Phaseolus vulgaris* L.), dolichos bean (*Dolichos lablab* L.), pea (*Pisum sativum* L.), bottle gourd (*Lagenaria siceraria* (Molina) Standl., cauliflower (*Brassica oleracea* L. var. *capitata*) on artificial inoculation in laboratory condition (Khatua *et al.*, 2014). Broccoli (*Brassica oleracea* L. var. *italica*) is also a host of this pathogen was reported by Kumar *et al.* (2003).

Disease management

a) As the pathogen has wide host range and do not have host specificity, care should be taken to remove and destroy diseased plant parts from the field by which survival potential of the pathogen can be reduced (Khatua *et al.*, 2014). Removal and destruction of infected plants from the field should be done before the formation of sclerotia. Infected fruits and other plant materials should not be kept for preparation of composting.

b) No resistant varieties are available, but plants with an open growth habit have less disease than plants with dense leaf coverage.

c) Wider row spacing is recommended for managing the disease.

2.5.1.18 Choanephora rot

Choanephora rot (also known as blossom blight, blossom end rot or wet rot), most commonly found on summer squash under wet conditions and sporadically on other cucurbits including pumpkin and vegetable marrow.

Symptoms

The disease is generally appears during flowering season and infection most commonly occurs on flowers, although the fungi can also infect through wounds on the fruit. Infected flowers are covered with white fungal growth initially which turn purplish black in time. In female flowers, the infection progresses into the fruit. The blossom end of the squash becomes soften, rotted and covered with fluffy purplish black fungal growth.

The causal pathogen and perpetuation

The disease is caused by the fungus *Choanephora cucurbitarum* (Berk. & Ravenel) Thaxt. The fungus survives from season to season in crop debris and is spread to new flowers by insects, splashing water or wind.

Disease management

a) Avoiding overhead irrigation the disease can be managed.

b) Wider spacing is helpful that provide adequate air movement in the field.

c) Raised plant beds and plastic mulch may be of help to limit fruit contact with moist soil and reduce moisture in the lower plant canopy.

d) Fungicides are ineffective against Choanephora rot because new susceptible flowers open every day. Besides, Copper hydroxide 77% WP @ 0.25% is effective against the disease.

2.5.1.19 Fusarium fruit rot of Pointed gourd

Fruit rot of pointed gourd incited by *Fusarium* sp. have been observed to be most serious problem in Ganga-diara area of Bihar, which causes 15% loss approximately. Nevertheless, the extent of affected fruits and fruit dropping due to *Fusarium* were invariably associated with varieties and different agro-climatic condition of Ganga-diara area (Khatua and Saha, 2004).

Symptoms

The first indications of the disease under field condition are water soaked patches on fruits. The skin of the infected portion becomes soft and succulent, resulting in disintegration of tissues. Slight swelling of skin around the infection court surrounds the brownish spots. In advanced stages of disease development, the fungus produces white growth on the yellowish brown affected region. The fruit looses turgidity; become yellowish and that infected fruit rot entirely without exudation. Discolouration of internal tissues and seeds occurs and the pulp turns into a light brown thick viscous fluid that gradually dry up. In certain cases, both end of the affected fruits become deep brown to black in colour. On splitting open, the affected fruits emits unpleasant odour (Khatua and Saha, 2004).

The causal pathogen and perpetuation

The causal pathogen is *Fusarium equiseti* (Corda) Sacc. Culture of the fungi, when first isolated are white with a floceose white mycelium tinged with peach but after 7-10 days it changes to beige and finally deep olive buff. From below the initial peach colour it changes to vinaceous fawn and finally dark brown. Only macroconidia are produced; these developed sparsely at first from simple lateral phialides but production increased with the formation of compact penicillately branched conidiophores after about 10 days. Conidia are falcate, with a well developed pedicellate foot cell and an attenuated apical cell which curves inwards. Mature conidia have 4-7 thin but distinct septa and measure 22-60 × 3.5-5.9μm. Chalamydospores are intercalary, solitary, in chains or knots, globose, 7-9μm in diameter (CMI description of pathogenic fungi and bacteria No. 571, 1978, Khatua and Saha, 2004).

Disease management

a) It is necessary to remove affected fruits from the field.

b) Spraying of fungicides used for controlling of fruit and vine rot disease will check the disease successfully (Khatua and Saha, 2004).

2.5.1.20 Fruit rot caused by other fungi

A large number of fungi were reported as causal agent of fruit rot of cucurbits due to their succulent nature and direct contact with soil. Any one of them may become serious in congenial condition on particular hosts. The fungi, *Choanephora cucurbitarum* (Berk. & Ravenel) Thaxt., *Diplodia natalensis Pole-Evans, Diplodia gossypina* Ellis & Everh. and *Mycosphaerella melonis* (Pass.) Chiu & Walker was reported on different hosts as inducing fruit rot symptom (Sohi and Sharma, 1998). Brown dull spots on ash gourd incited by *Fusarium solani* (Mart.) Appel & Wr. and *F. moniliforme* Sheld. was observed by Upadhyay and Roy (1987). Vyas and Panwar (1976) recorded small, circular, olive green patches caused by *Myrothecium roridum* Tode, and light brown, water soaked spots caused by *Rhizoctonia solani* Kuhn on bitter gourd. On cucumber, Laxminarayana and Reddy (1976) recorded different types of characteristic symptoms caused by several fungal plant pathogens viz. *Rhizoctonia solani* (discoloured, water soaked spots), *Colletotrichum capsici* (Syd.) Butler & Bisby (irregular discolouration), *Helminthosporium hawaiiense* Bugnic., (pale brown depression), *Curvularia pallescens* Boedijn (style end rot, circular, brown spots), *Alternaria tenuis* Nees (round, pale brown lesions) and *Myrothecium roridum* (dark brown patches). Circular, necrotic lesions on muskmelon incited by *Myrothecium roridum* was reported by Singh (1986). Gangopadhyay and Sharma (1976) recorded drying of fruit followed by rotting of rind and sponginess of pulp on pumpkin due to the infection of *Rhizoctonia solani* and *Fusarium oxysporum* (Schlecht). Snyder & Hansen. Siddaramaiah *et al.* (1982) found that the fungi *Sclerotium rolfsii* is responsible for causing soft rot of pumpkin that develop water soaked areas on the fruits; whereas, Dalela (1956) recorded *Rhizopus* sp. as soft rotting fungi of pumpkin. Rangaswami and Mahadevan (2004) also mentioned *Rhizopus* spp. and *Erwinia* spp. as dry and soft rot causing fungi, respectively. Laxminarayana and Reddy (1976) recorded *Sclerotium rolfsii* (soft rot), *Alternaria tenuis* (brown spots) and *Phoma* sp. (circular spots) on ridge gourd. On water melon, water soaked, brown discolouration incited by *Cladosporium tenuissimum* Cooke was reported by Narain *et al.* (1985). *Fusarium equiseti* produces brown dull spots on sponge gourd (Prasad and Ambasta, 1987). But, *Fusarium oxysporum* is responsible for producing similar type symptom on tinda (Mathur and Mathur, 1958, Kore and Kharwade, 1987).

2.5.2 Diseases caused by Bacteria

2.5.2.1 Angular Leaf spot

This bacterial disease primarily affects cucumber, but it may occur on muskmelon, squash, pumpkin, and watermelon. In cooler regions, this disease is very common and destructive. Mukherjee and Khatua (1998) reported that the disease is common in West Bengal condition.

Symptoms

The disease occurs on leaves, young green stems, and also on fruits. The disease appears on the leaves as small, yellowish water-soaked spots.

The shape of older lesions tends to be irregular or angular (mostly triangular) as they enlarge and encounter veins. Under hot and humid conditions, whitish bacterial ooze may be found on the undersides of lesions. It dries to form white crusts. The spots enlarge and become brown. Afterward, the tissues dry and may fall out leaving irregular holes in the leaf. Water-soaked necrotic spots may also be developed on stems and fruits. On fruits, the spots are much smaller and nearly circular. When the diseased portion dies, the tissues become white and crack. The fruit lesions are generally superficial but cracking may expose it to the secondary attack of soft rot bacteria. The bacteria may move deep into the fruit and infect the seeds.

The causal pathogen and perpetuation

Angular leaf spot is caused by *Pseudomonas syringae* pv. *lachrymans* (Smith & Bryan) Young *et al.*. This bacterium can overwinter in seed and on infected plant parts left in the field. The pathogen can survive in plant debris for over 2 years. The disease can be introduced into a field through contaminated seed. The pathogen is disseminated by splashing rain, wind-blown rain, wind-blown soil, insects, farm equipment, and field workers. Infection occurs through natural openings and wounds. The disease is favored by warm temperatures and high humidity.

Disease management

a) Pathogen-free seed should be used to avoid the disease.

b) Hot water treatment can be done for cucumber seed.

c) Crop rotation with non-hosts for at least 2 years is helpful for managing the disease.

d) Management of irrigation systems is important to minimize leaf wetness and soil splash. Overhead irrigation should be avoided.

e) Proper ventilation facility is important in managing this disease in greenhouse production.

f) Use of resistant varieties is important to avoid the disease.

g) Application of copper based fungicides (Copper oxychloride 50% WP @ 0.4% or Copper hydroxide 77% WP @ 0.25%) may help protect plants when conditions are favorable for the disease. However, Copper based fungicides are not effective if epidemic have already been started. Spraying of Streptocycline 90:10 SP @ 100ppm reduce the spread of the disease in the field.

2.5.2.2 Bacterial wilt

Bacterial wilt of cucurbits caused by *Erwinia tracheiphyla* (Smith) Bergey *et al.* is a serious disease occurs in different parts of the world including United States, Europe, South Africa and Japan. In India, the disease is common in Haryana, Punjab, West Bengal, Assam, Orissa and Maharashtra (Mustafee, 1998). Incidence of the disease in sweet gourd, watermelon and bottle gourd is rare in West Bengal condition (Mukherjee and Khatua, 1998). Many species of the family cucurbitaceae i.e. cucumber, muskmelon, squash and pumpkin are reported to be infected by the disease. This pathogen can cause severe losses in cucumbers and muskmelons; squash and pumpkins are less severely affected. Watermelon is not affected by the disease.

Bacterial wilt of bottlegourd (*Lagenaria siceraria* (Molina) Standl caused by a gram negative bacterium, *Ralstonia solanacearum* (Smith) Yabuuchi *et al.* was also recorded from West Bengal (India) by Mondal *et al.* (2004, 2011 and 2014) with very lower disease incidence (4.50%).

Symptoms

The disease (caused by *Erwinia tracheiphyla*) produces two types symptoms: i) Sudden wilting of foliage and vines of the affected plant in field condition ii) Slime rot of squash fruit in storage condition.

In the first case, symptom appear as drooping of one or more leaves of the vines with dull green appearance and consequently, the infected plant shows wilting of all leaves and collapse of all vines. Wilted leaves of the affected plant shrink and dry up; affected stems primarily become soft and pale, but later shrink and become hard and dry. Wilt progresses down the vine until entire vine is wilted. In moderately resistant plants or under adverse situation, symptom progress slowly and may rarely be accompanied by excessive blooming and branching of the infected plants. Wilt progression varies by crop. Cucumber and melon wilt and die rapidly whereas pumpkin takes up to two weeks to wilt completely. Summer squash may continue to produce wilting symptom several weeks after infection.

If infected stems are cut and pressed between the fingers, droplets of white bacterial ooze appear on the cut surface. If infected vines are cut close to the crown of the plant and the cross sections pressed together, thread like strands of bacterial ooze can be seen when the two halves are gently pulled apart again. Measures have to be taken observing the presence of striped or spotted cucumber beetles in the field that are actually acts as vector.

In case of slime rot of stored squash fruit, the disease progress internally whereas the outer surface of the fruit may appear normal. In time, dark spots or blotches appear on the surface of the squash, which later coalesce and enlarge. The disease develops over several months in storage. These squash fruits may be infected secondarily by other soft-rotting microorganisms and are smashed entirely.

The causal pathogen and perpetuation

The disease caused by *Erwinia tracheiphila* is a gram negative, rod shaped, peritrichous bacterium, survives for only a few weeks in infected plant debris. The bacterial pathogen may also survive in weeds without producing any visible symptoms. The bacterium overwinter (hybernates) in the intestines of striped cucumber beetles, *Acalymma vittata* (F.) and spotted cucumber beetles, *Diabrotica undecimpunctata* Mannerheim. These two beetles are also responsible for transmission of the bacterium.

During spring, the cucumber beetles that carry bacteria feed and cause leaves injury and deposit bacteria in the injury with their feaces. The bacteria enter the xylem vessels, multiply rapidly, and spread to different parts of the plant. Obstruction of the xylem vessels of infected plant occurs due to rapid multiplication of the bacteria and their polysaccharides, deposition of gum and formation of tyloses in the xylem elements, which allow less than one-fifth of the normal water flow. This extensive plugging of the vessels is the primary cause of wilting.

The bacteria are spread to the healthy plants through contaminated mouthparts of different insects particularly by the striped and spotted cucumber beetles. After one feeding on a wilted plant each contaminated insect can infects several healthy plants though a relatively small percentage of beetles become carriers of the bacteria. The first wilting symptom appears on the field at 6 to 7 days after infection and another 7 days is required to wilt the plant completely. The bacteria present in the xylem vessels of infected plants die within 30 to 60 days after the dead plants dry up.

Fruit infection of squash plants generally occurs through infected vines and rarely through beetles feeding on the blossoms and the rind of developing squash.

Disease management

a) If disease appears in a few plants, rogueing and burning of these plants is helpful to prevent further spread of the disease.

b) Monitoring of cucumber beetles during growing season followed by proper management practices through insecticides and others provides the most effective control of bacterial wilt (Kennelly, 2012).

c) Bactericides/antibiotic or other pesticides will not facilitate in managing cucurbit plants infected with this destructive bacterial disease.

d) To circumvent squash rot in storage, fruits only from healthy plants should be picked and it should be stored in a clean, fumigated warehouse.

e) Cultivation of resistant cucurbit varieties is most important to avoid the disease. Varieties with less cucurbitacin show less damage. Presence of cucurbitacins stimulates cucumber beetles and rootworms to feed on cucurbits (Hoffman and Zitter, 1994).

f) Attractant-baited traps and attracticidal baits are promising for control of cucumber beetles. Attracticidal baits are mixtures of a feeding stimulant (cucurbitacins) and a small amount of insecticide; some baits also include a mixture of volatile attractants. In theory, the cucurbitacins stimulate the beetles to feed, and the volatile attractant draws them to the bait. The biggest advantage of attracticidal baits is that the amount applied per acre is very small relative to a standard foliar application. Traps being developed could be used for early detection of infestations and possibly for control if several traps are used per field (Hoffman and Zitter, 1994).

g) Only a few species of parasitoids (tachinid fly and braconid wasp) or insect pathogens have been observed affecting cucumber beetles and their impact is not well documented. The larvae are attacked by predacious nematodes. One species of soldier beetle is considered an important predator (Hoffman and Zitter, 1994).

2.5.2.3 Bacterial leaf spot

Bacterial leaf spot is a serious seed borne disease causes considerable damage to the cucumber crop.

Symptoms

The disease primarily appears as small, water soaked areas on the under surface of the leaves. The upper surface of these areas of leaves looks yellow. The spots are inter-veinal in nature. In time, the spots emerge and become angular. The colour of the spots changes to brown. The spots are surrounded by a chlorotic halo. Several spots may coalesce to form larger spots. Afterward, the tissues dry. Occasionally young stems and petioles are also attacked. On stems and petioles, the lesions are linear or streak like and brown in colour.

The causal pathogen

The disease is caused by a gram negative bacterium, *Xanthomonas cucurbitae* (ex Bryan) Vauterin *et al.* (=*Xanthomonas compestris* var. *cucurbitae* (Bryan) Dye).

Disease management

a) To kill the bacterium in seeds, hot water treatment (50°C for 30 minutes) is recommended. Care should be taken for this practice; otherwise germination percentage of the seed may be reduced.

b) It is better to obtain seed grown in dry areas.

c) Spraying of copper fungicides (Copper oxychloride 50% WP @ 0.4% or Copper hydroxide 77% WP @ 0.25%) or Streptocycline 90:10 SP @ 100ppm reduce spread of the disease in the field.

2.5.3 Viral Diseases

The commonly grown cucurbits *viz.* pumpkin, cucumber, bitter gourd, ridge gourd, sponge gourd, bottle gourd, ash gourd, water melon and pointed gourd are naturally infected by more than 10 viruses. The most common ones are cucumber green mottle mosaic virus (CGMMV), cucumber mosaic virus (CMV), water melon mosaic virus (WMV) and numbers of different gemini virus groups. Besides, some unidentified viruses also infect cucurbits. A few cucurbits are also affected by phytoplasma diseases, which are however economically not very important. Recently, the incidence of viral diseases is increasing in trend. Earlier, the white fly transmitted viral diseases were of minor importance, while in recent time they are appearing in almost epidemic proportion (Varma and Giri, 1998).

Almost all the cucubitaceous vegetables are subjected to attack mosaic virus affecting leaf and fruits (Chowdhury *et al.*, 1998). According to Nath (2004) cucurbits are mainly suffer from cucumber mosaic virus (CMV) and water melon mosaic virus (WMV). Mukhopadhyay and Saha (1968) observed that CMV could be transmitted through seeds of *Cucurbita moschata* Duchesne ex Poir., whereas sap transmission of WMV was observed by Ghosh & Mukhopadhyay (1979). Ghosh and Mukhopadhyay (1977a,b) described eighteen different viruses (bottlegourd mosaic, vein yellowing, cucurbit latent, watermelon mosaic, pumpkin enation mosaic, pumpkin mild mosaic and a strain of squirting cucumber mosaic virus) infecting pumpkins. They also estimated the biochemical differences between the pumpkin plants infected by these viruses particularly buffer soluble proteins, reducing sugar and starch (Mukhopadhyay and Nath, 1998). Bandopadhyay and Mukhopadhyay (1977) described a strain, cucurbit latent virus infecting muskmelon. Sarkar and Mukhopadhyay (1979) separated nine anisometric viruses of cucurbits by host reactions. Mukhopadhyay (1988) recorded Arka Chandan as resistant variety against cucurbit latent virus and WMV. He identified Midnapore local as a resistant source only to bottle gourd mosaic virus. Pointed gourd crop is affected by CMV and particle of the virus is isometric in shape (Nath, 2004). Bitter gourd is infected by Zucchini yellow mosaic virus and *Aphis gossypii* Glover acts as vector whereas ridge gourd is affected by several viruses (Nath, 2004). Host range of the Zucchini yellow mosaic virus and papaya ring spot virus is restricted within cucurbits whereas watermelon mosaic virus not only infects cucurbits but also some weeds, and is transmitted by aphids. Aphids can acquire then transmit the virus within a few second. Once acquire, they lose it after a few plants. The squash mosaic virus infects cucurbits alongwith some other hosts under the family chenopodiaceae, which is reported to be transmitted by seeds and cucumber beetles (Kennelly, 2012). Virus nature of the disease was confirmed through ISEM study and the particles of the viruses were found to be flexious rod, and the viruses were under poty virus (Nath, 2004). Mukhopadhyay and Nath (1998) confirmed the prevalence of WMV in *Cucurbita moschata, Lufa acutangulata* and *Coccinia* sp., CMV in *Cucumis sativa, Trichosanthes dioica* and *Cleome viscose,* and Zucchini yellow mosaic virus in *Lufa acutangulata* and *Coccinia* sp. Different viruses infecting cucurbitaceous crops in West Bengal and their key characters are mentioned below (Mukhopadhyaya and Nath, 1998).

Disease	Virus group	Key characters
Cucumber mosaic	Cucumovirus	Isometric ssRNA aphid-transmitted
Squash mosaic	Comovirus	Isometric ssRNA beetle-transmitted
Cucumber green mottle	Tobamovirus	Anisometric ssRNA not known vector
Watermelon mosaic	Potyvirus	Anisometric ssRNA aphid-transmitted
Muskmelon vein mosaic	-	Anisometric ssRNA aphid-transmitted
Yellow vein mosaic	-	Whitefly transmitted
Cucumber stunt mottle	Nepovirus	Isometric ssRNA nematode-transmitted
Cucumber necrosis	Tobacco Necrosis virus	Isometric ssRNA fungus-transmitted
Melon mosaic	-	-

2.5.3.1 Mosaic diseases

A variety of mosaic symptoms occur on different members of cucurbitaceae. In India, these are common on almost all cucurbits and often cause serious losses. The common mosaic is caused by the cucumber mosaic virus. In addition, cucumber green mottle mosaic virus and watermelon mosaic virus are also common in certain cucurbits and cause significant losses. Most of the cucurbits viz. cucumber, melon, snap gourd, long melon, pumpkin, water melon, round gourd, bottle gourd, ridge gourd, sponge gourd, bitter gourd and snake gourd, grown in different parts of India are affected by viras diseases. Infected plants produced varied types of symptoms, which include (a) mild or severe dark green mosaic (b) yellow-green or bright yellow mosaic mottling of the leaves, accompanied by blistering of green areas. In these cases plants produces variously deformed leaves. Stunting of growth and reduction of leaves size are the common phenomenon that ultimately affects the yield (Rani *et al.*, 1971, Cheema *et al.*, 1999).

a) Cucumber Mosaic Virus (Cucumovirus)

Cucumber Mosaic Virus (CMV) is a most widely distributed plant virus that has a very wide host range (Gour *et al.*, 2008). The virus infects about 200 dicotyledonous and monocotyledonous plants belonging to about 40 families (Franski and Hatta, 1980). According to Kennelly (2012) the virus infects more than 800 plant species.

Symptoms

Symptoms primarily appear on younger leaves which curl downward and become mottled, distorded, wrinkled and reduced size. The growth of the plant is stunted. Internodes of the infected plant become shorter. Rosetting of the youngest leaves is common. Fruits produced by such plant are often mis-shaped, mottled, warty and reduced in size. Infected fruit may look like a 'white pickle' in which there is very little green colour. If a plant becomes infected after midseason, vine growth may not be reduced, but developing fruit may be bumpy and deformed (Gour *et al.*, 2008).

The causal pathogen, its transmission and epidemiology

The virus is characterized by isometric particles measuring 28-30nm in diameter (Shankar *et al.*, 1971, Giri, 1985). Some strains of the virus in India possess a satellite RNA that attenuates system development (Giri, 1985). *Datura stramonium* L. is a good diagnostic and *Chenopodium amaranticolor* (Coste & A.Reyn.) Coste & A.Reyn. a good assay host (Varma and Giri, 1998). The particles appear spherical with smooth outline and darkly stained central region due to the penetration of negative stain (Varma and Giri, 1998). This virus is efficiently transmitted in a nonpersistent manner by different species of aphids viz. *Aphis craccivora* Koch., *A. gossypii* Glover, and *Myzus persicae* Sulzer (Bhargava and Bhargava, 1977, Giri, 1985, Joseph and Ramanath, 1978, Joshi, 1977, Rao, 1976, Raychaudhuri and Varma, 1975a, Singh *et al.*, 1976, Tripathi and Joshi, 1985, Gour *et al.*, 2008, Kennelly, 2012). The virus is known to be transmitted through the seeds of several species of cucurbits, but in India, it is reported to be seed transmitted in only pumpkin and vegetable marrow (Mukhopadhyay and Saha, 1968, Sharma and Chohan, 1973). According to Gour *et al.* (2008) the virus survives on weeds, ornamentals and other crops. Pickers may serve to spread the virus from one plant to another during harvesting.

The CMV causes mosaic disease in cucumber (Uppal, 1934, Bhargava and Bhargava, 1977), muskmelon (Mayee *et al.*, 1976, Sharma *et al.*, 1984), pumpkin, summer squash and vegetable marrow (Reddy and Nariani, 1963, Rao, 1976, Bhargava and Bhargava, 1977), snake gourd (Pillai, 1971, Joseph and Ramanath, 1978) and watermelon (Bhargava and Bhargava, 1977) in different parts of India. The disease rarely affects watermelon (Gour *et al.*, 2008).

b) Cucumber Green Mottle Mosaic Virus (CGMMV) (Tobamovirus)

Cucumber green mottle mosaic virus (CGMMV) was first reported from Great Britain in *Cucumis sativus* by Ainsworth in 1935 (Okada, 1986). In India, the disease was reported first time in bottle gourd (Capoor and Verma, 1948, Vasudeva *et al.*, 1960, Mandal *et al.*, 2008). CGMMV is distributed in Europe and Asia (Eurasian region), and it is one of the few plant viruses that has been recorded in Antarctica (Polischuk *et al.*, 2007, Mandal *et al.*, 2008). It has rod shaped particles measuring 300 × 15nm (Shankar *et al.*, 1971, Raychaudhuri and Varma, 1978). The particles sediment as a major component at 192 S and two minor components at 104 S and 76 S. They contain approximately 5 per cent nucleic acid (Raychaudhuri and Varma, 1978). The particles in leaf dip preparation can be grouped into three classes, indicating easy breakage of particles measure 158nm in length. They are larger than those found in some other isolates of CGMMV (Fukuda *et al.*, 1981). The virus is a strong immunogen and is distantly related serologically to Franzipani mosaic and TMV (Vani and Varma, 1988).

Several cucurbitaceous vegetables such as bottle gourd (*Lagenaria siceraria*), cucumber (*Cucumis sativus*), gherkin (*Cucumis anguria*), muskmelon (*Cucumis melo*) and watermelon (*Citrullus vulgaris*) are affected by the CGMMV (Mandal *et al.*, 2008, Nagendran *et al.*, 2015). High incidence of CGMMV i.e. 100% in bottle gourd, 80% in muskmelon, 75% in watermelon has been recorded in North Indian conditions (Raychaudhuri and Varma, 1978, Rao and Varma, 1984). A loss of 10 to 15% was reported due to CGMMV infection in cucumber grown in greenhouses in Canada by Ling and his associates in 2014 (Nagendran *et al.*, 2015) and in China (Shang *et al.*, 2011). Serious outbreak of CGMMV in cucumber and watermelon is known as early as 1966-1969 in Japan (Komuro, 1971). Subsequently, the disease appeared as a serious problem in different parts of the world i.e. in 1995 in Greece, 2002 in Korea and 2005-06 in China (Bem and Vassilakos, 2000, Yoon *et al.*, 2008, Chen *et al.*, 2008). CGMMV being a potential threat to the production of cucurbitaceous crops has been recognized as a quarantine pest by the Government of China in May 2007 (Chen *et al.*, 2008, Mandal *et al.*, 2008).

Symptoms

The typical symptoms caused by CGMMV are systemic greenish mottle mosaic on foliage, however, the nature of the symptoms and loss depends upon the time of infection, type of host and the strain of the viruses associated with the disease. Early infection produces aggressive symptoms and causes severe losses. When the plant is attacked soon after emergence, the cotyledons are yellow and seedlings show symptoms of wilt. In case of older plants, symptoms appear first on younger leaves showing alternate green and yellow patches, which are of irregular in shape and enlarge rapidly, ultimately cover the entire leaf. The diseased leaves are mottled, deformed, small and sometimes curled downward. The veins and veinlets also turn yellow. Sometimes, there are shallow depressions on the leaves. Shortening of the internodes followed by dwarfing of the plant are common. The activity of the virus is reduced when the leaf ages. Young fruits are rough, mottled and deformed. Often these fruits are white and much smaller in size than normal ones. Fruit drop is also common. Yield is reduced by 25% even more if no management options are taken into account. In the epidermis of the infected plants characteristic structures can be seen easily with a microscope.

CGMMV causes serious distortion and decomposition of fruits. In grafted watermelon plant it causes 'Blood Flesh' disease in Korea (Lee *et al.*, 1990), 'Konnyaku' disease in Japan (Komuro *et al.*, 1968).

Infected cucumber, bottle gourd, and cantaloupe plants consisted of young fully expanded leaves showed symptoms of mottling, systemic mosaic, mild to severe yellow mottling, vein clearing, blistering, and growth stunting (Moradi and Jafarpour, 2011). The acuba mosaic strain of CGMMV causes yellow mottling on leaf and yellow streak or fleck on fruits of cucumber.

On snake gourd symptoms appear as mosaic, mottling with reduction of leaf size, and phylloid flowers (Nagendran *et al.,* 2015).

The causal pathogen, its transmission and epidemiology

The virion is rod shaped particle of 300 + 15 nm containing positive sense genome. CGMMV is highly stable and contagious. Natural spread of the virus is largely through contact of infected plant materials. The virus is efficiently transmitted through mechanical sap inoculation. The virus does not have any specific vector, although cucurbit leaf beetle *Aulacophora foveicollis* Lucas is able to transmit the virus to the extent of 27 per cent (Rao and Verma, 1984, Vani and Varma, 1988). Infective virus particles were detected in the regurgitated fluids and excreta of the leaf beetle feeding on the diseased leaves. This is not a seed transmitted virus disease in muskmelon, vegetable marrow and bottle gourd (Rao and Varma, 1984, Vani and Varma, 1988). However, in Japan and the Netherlands, up to 8 per cent seed transmission of the virus was reported in water melon, cucumber and bottle gourd (Kamuro *et al.,* 1971, Van Koot and Van Dorst, 1959). Commercially important watermelon varieties are generally cultivated by grafting on the rootstoch of bottle gourd or other cucurbits in China, Greece, Japan and Korea, where seed and graft transmission have been attributed to introduction and emergence of CGMMV (Mandal *et al.,* 2008). Under field conditions, the virus is transmitted through contaminated tools, but not through soil or pollen. However, under experimental condition, 18% transmission of CGMMV has been demonstrated in bottle gourd when soil was mixed with freshly dried infected plant debris. The infected plant debris possibly contributed as primary source of infection to a limited number of plants and successively the virus spreads through plant to plant contact (Rao and Varma, 1984). Irrigation water plays an important role in the spread of the virus (Vani and Varma, 1988). The virus remains infective in the contaminated soil for at least 10 months (Varveri *et al.,* 2002). The cutting knives used by growers for harvesting fruits from diseased plants can potentially contribute additional means of field spread of CGMMV diring fruiting stage of crop. Under experimental condition, the CGMMV could be transmitted up to four healthy bottle gourd plants through serial cuttings by contaminated razor blade (Mandal *et al.,* 2008).

The host range of the virus is restricted to the members of Cucurbitaceae and Chenopodiaceae. *Chenopodium amaranticolor* is a good assay host (Varma and Giri, 1998).

The virus was found to naturally infect bottle gourd, musk melon and water melon in India (Shankar and Nariani, 1974, Raychaudhuri and Varma, 1975a) whereas in other countries, it causes disease in cucumber. It is interesting that even though the virus infects *Cucurbita pepo* L. easily by mechanical inoculation, no natural infection occurs in this host (Varma and Giri, 1998). Nagendran *et.al.* (2015) also reported mechanical transmission of the virus by inoculating *Nicotiana glutinosa* plants with crude sap extracts from symptomatic leaves of snake gourd. They first documented serological and molecular evidence for the occurrence of CGMMV in snake gourd in India.

Other tobamovirus infecting cucurbits are Cucumber fruit mottle mosaic virus (CFMMV), Kyuri green mottle mosaic virus (KGMMV), Zucchini green mottle mosaic virus (ZGMMV) was reported by the scientists from different parts of the world (Antignus *et al.*, 2001, Francki *et al.*, 1986, Yoon *et al.*, 2001, Ryu *et al.*, 2000, Yoon *et al.*, 2002).

c) Papaya Ring Spot Virus (PRSV-W), Water Melon Mosaic Virus-2 (WMV-2) and Water Melon Mosaic Virus (WMV) (Potyviruses)

Papaya Ring Spot Virus (PRSV-W), Water Melon Mosaic Virus-2 (WMV-2) and Water Melon Mosaic Virus (WMV) are amongst the top 10 economically significant filamentous viruses of the Indian subcontinent (Varma, 1988, Varma and Giri, 1998). Earlier, they were considered strains of the watermelon mosaic virus. But, now they appeared to be distinct on the basis of host range study and antigenicity test (Varma and Giri, 1998). In India, the earliest record of this virus seems to be that of Vasudeva and Lal (1943), who described it as bottle gourd mosaic virus causing mosaic disease. It was difficult to distinctively identify the virus, because of the studies of Vasudeva and Lal (1943) were limited only on host range and physical properties. Later investigations revealed that PRSV-W and WMV-2 are anisomatric filamentous viruses with particle length of 720-760 nm (Raychaudhuri and Varma, 1975a). Slight variations in the length of particles were found in different isolates of the virus (Bhargava, 1977).

Symptoms produced by watermelon mosaic virus

Watermelon mosaic virus and Watermelon mosaic virus-2 infect all cucurbits, and are produced mosaic symptoms. Infected leaves first develop a yellowing between the veins. Afterwords the leaves become mis-shapen. Except the tissues, adjacent to veins and veinlets, remaining lamina surface may be destroyed. The vein and veinlets frequently protrude beyond the leaf margin and the leaf tissue around the major veins may develop a tendril-like appearance (filiform or spindle shaped). The vein in these infected leaves originate from the base of the leaf stalks. New leaves are mottled, blistered and distorted. Infected plants show stunted in growth. Poor flowering and defoliation are the other defects. Fruits can become bumpy and severely distorted with occasional changes in colour. This virus is the most dangerous among the entire cucurbit virus (Varma and Giri, 1998, Gour *et al.,* 2008).

The causal pathogen, its transmission and epidemiology

The viruses, PRSV-W and WMV-2 are easily transmitted by sap inoculation, but no evidence of seed transmission was recorded in India (Raychaudhuri and Varma, 1975a). These viruses are efficiently transmitted by different species of aphids in a nonpersistent manner (Shankar *et al.,* 1969, Raychaudhuri and Varma, 1975b, Darekar and Sawant, 1989). Although, the aphids viz. *Aphis gossypii, A. craccivora, Macrosiphum sonchi* L., *Rhopalosiphum maidis* Fitch and *Myzus persicae* transmit PRSV-W in India (Bhargava *et al.,* 1975), while WMV-2 transmitted efficiently through *A. gossypii* and *Myzus persicae* (Raychaudhuri and Varma, 1975a, 1975b). *Aphis gossypii* is the most important vector for transmission of these viruses (Bhargava *et al.,* 1975, Varma and Giri, 1998).

Isolates of Papaya Ring Spot Virus (PRSV) from cucurbits resemble the type W strain of the virus (Purcifull *et al.,* 1984). The host range of all the isolates of PSRV-W is restricted within the family cucurbitaceae. The virus causes mosaic diseases in *Cucurbita maxima* Duch. (Singh, 1981), *C. moschata* Duch. ex Poir. (Ghosh and Mukhopadhyay, 1979), *C. pepo* (Reddy and Nariani, 1963), and vegetable marrow (Raychaudhuri and Varma, 1975a). The important reservoirs of PRSV-W are vegetatively propagated *Trichosanthes dioica, Lagenaria vulgaris, Momordica dioica* and *Coccinia grandis* (Bhargava *et al.,* 1975). Water Melon Mosaic Virus (WMV-2) has a wide host range, causes severe disease in *C. pepo* and other cucurbits (Raychaudhuri and Varma, 1975a, 1975b).

Bottle gourd, chow-chow, pointed gourd, water melon and ash gourd were reported to be infected with Water Melon Mosaic Virus (WMV) from different parts of India. The virus produces mosaic disease and sometimes causes severe damage to the crops (Shankar *et al.,* 1972, Bhargava and Bhargava, 1977, Singh, 1981, Vani, 1987). The disease is more severe in bottle gourd when the plants are infected with both Cucumber Green Mottle Mosaic Virus (CGMMV) and WMV. Mahmood *et al.* (1974) reported that the disease in bottle gourd is more severe when the plants are infected also with powdery mildew disease (c. o.: *Sphaerotheca fuliginae*). Susceptibility of the plants to root knot nematode (*Meloidogyne incognita*) may be increased due to the infection of WMV (Mahmood *et al.*, 1974, Nayar and More, 1998). WMV is a serious problem of chow-chow in Karnataka (Singh, 1981). The virus is efficiently transmitted by aphids and through seeds. In water melon, mosaic disease starts appearing about 40 days after sowing and reaches about 80 per cent incidence by the end of the season in Delhi condition (Vani, 1987). Bhargava and Bhargava (1977) reported the WMV causing mosaic disease of ash gourd and pointed gourd from Uttar Pradesh, which is sap transmissible and also transmitted by *Myzus persicae* and *Aphis gossypii* in a nonpersistent manner. In case of ash gourd, the host range of the virus is restricted to cucurbits and *Vigna sinensis* whereas, in pointed gourd, the host range is restricted to cucurbits, *Vigna sinensis* and *Zinnia elegans.* Aphids and in some cases leaf miners can transmit the watermelon mosaic virus. The virus spreads by farm machinery and pickers, as well as by insects. The disease is severe during worm growing season (Gour *et al.*, 2008).

Roy and Mukhopadhyay (1979) studied the epidemiology of watermelon virus-1 (WMV-1) in pumpkin and detected the virus in root, leaf debris and seeds. They not only recorded 86% transmission by *Aphis gosypii* but also observed transmission of the virus through contact between the above ground parts of the plants. Manual handling is another method of transmission of the virus was recorded by them (Mukhopadhyay and Nath, 1998).

2.5.3.2 Other Viral diseases

a) Geminiviruses

Geminiviruses was found associated with bitter gourd, musk melon, pumpkin, sponge gourd (Giri *et al.*, 1981) and cucumber (Varma, 1955, Ghosh and Mukhopadhyay, 1979a) from different parts of India. The virus causes leaf distortion in bitter gourd and yellow vain in cucumber and pumpkin, and incidence of the disease in bitter gourd recorded nearly 80 per cent (Giri *et al.*, 1981, Varma, 1955, Ghosh and Mukhopadhyay, 1979). Leaf distortion in bitter gourd is very common in northern and eastern India during rainy season. Besides, the diseased plants develop veinal chlorosis and chlorotic spots. The internodes of infected plants are shortened and axillary buds proliferate. Fruiting is reduced. Fruits are deformed and contain shriveled seeds. Infection results in about 18 per cent reduction in viability of pollen. The symptoms are very prominent in the susceptible variety, Faizabadi (Giri *et al.*, 1981, Varma and Giri, 1998). In India, yellow vein of cucumber was first reported from Pune (Varma, 1955), and later from West Bengal (Ghosh and Mukhopadhyay, 1979). In 1989, the disease of cucumber and pumpkin appeared in epidemic form in Delhi (Varma and Giri, 1998). Yellow vein disease of pumpkin was reported from Maharashtra (Capoor and Ahmad, 1975), Uttar Pradesh (Bhargava and Bhargava, 1977) and West Bengal (Ghosh and Mukhopadhyay, 1979). The viruses of this group geminate into particles measuring 30 × 19 nm formed by the joining of two icosahedral particles. The concentration of the particles is greater in fully expanded young leaves than in the very young and old leaves. These viruses are efficiently trapped by the antiserum to squash leaf curl virus in immune-electron microscopy tests. Various geminiviruses can be distinguished by their reactions to panels of monoclonal antibodies to African and Indian cassava mosaic viruses (Varma and Giri, 1998).

The geminivirus infecting bitter gourd is easily graft-transmissible. The virus is not transmitted by sap inoculation, whereas it is efficiently transmitted by the insect vector, whitefly (*Bemisia tabaci*), in a persistent manner. The virus is also seed-borne in bitter gourd (Mishra *et al.*, 1983).

The viruses of this group are gradually becoming more prevalent and they may become a serious problem for cucurbit cultivation. The ecological conditions that favour cucurbit cultivation also favour these viruses (Varma and Giri, 1998).

b) Tobacco Ring Spot Virus (Nepovirus)

Nepovirus was found associated with musk melon (*Cucumis melo*) and long melon (*Cucumis melo* cv. *utilissimus*) with different types of symptoms viz. mosaic, ring spot etc.

Ring spot type symptoms of musk melon were reported by Vishwanath *et al.* (1980) from Delhi. In ring spot, the leaves of diseased plants develop mosaic or mottling with chlorotic areas that subsequently form a ring spot pattern. Occasionally, young leaves develop small yellowish brown dots surrounded by bright yellow halo or margin (Vishwanath *et al.*, 1980). The causal virus disease is transmitted mechanically, but not by aphids. It also infects *Nicotiana tabacum, N. glutionosa* and *Petunia hybrid,* producing systemic symptoms, and induces local lesions on inoculated leaves of *Cucumis amaranticolor.* The virus has been tentatively identified as tobacco ring spot virus (nepovirus) (Varma and Giri, 1998).

Mosaic type symptoms of long melon were reported and described by Rani *et al.* (1971) and Raychaudhuri (1973) from Delhi. They recorded 70 per cent disease incidence. The leaves of the diseased plants show fine vein clearing followed by mottling and well defined mosaic. The mottling usually starts on the margins of leaves and gradually progresses towards the petiole. Leaf size is reduced without any apparent blistering or puckering. The virus is not transmitted by aphids and seeds and it infects mostly cucurbit hosts. *Chenopodium amaranticolor* is a good assay host and *Datura stramonium* is a diagnostic host (in which the virus produces severe symptoms). On the basis of serology, Rani *et al.* (1971) identified the Lucknow isolate to be tobacco ring spot virus. However, this needs further confirmation as various properties of the causal virus suggest that it may belong to comovirus group (Varma and Giri, 1998).

c) Cucumber Latent Virus (CLV)

A latent virus disease of muskmelon was reported from West Bengal with 25-40 per cent incidence (Bandyopadhyay and Mukhopadhyay, 1977). The causal virus was tentatively identified as a strain of cucumber latent virus (Webb and Bohn, 1961, Webb, 1963). The virus is transmitted by sap inoculation and by the aphids, *Aphis gossypii, A. craccivora* and *Myzus persicae* in a nonpersistent manner. The virus is transmissible to *Cucumis sativus, Cucurbita moschata, Luffa acutangula, Vigna sinensis* and *Cucumis amaranticolor*. The virus is not transmissible to *Nicotiana tabacum* cv. White Burley and *Datura stramonium.* It is revealed from the properties of the virus that it may be a strain of CMV (Varma and Giri, 1998).

Plumb *et al.* (2000) recorded different viruses of crops and weeds in Eastern India. They presented a clear picture about the following viral diseases of cucurbitaceous vegetables. The detail of cucurbits infecting viruses, their particle characteristics and physical properties (Smith, 1972, Narayanasamy, 2001) are described below.

Table 6. Viral diseases of pumpkin in Eastern India

Particle characteristics and physical properties of viruses	Transmission and interseasonal carryover
Watermelon mosaic virus 1 (WMV 1), the unenveloped filamentous to flexuous *Potyvirus* with a modal length of 760-800 nm x 12 nm, (+)ss RNA, DEP 1:10000-1:30000, TIP 55-60°C, LIV 9-10days.	Transmit mechanically and by aphid vector *i.e. Myzus persicae, Aphis craccivora* and *Macrosiphum euphorbiae* in non-persistant manner. Interseasonal carryover occurs through infected crops and weeds.
Watermelon mosaic virus 2 (WMV 2), the unenveloped filamentous to flexuous *Potyvirus* with a modal length of 750 nm x 12 nm, (+)ssRNA, DEP 1:10000-1:30000, TIP 55-60°C, LIV 9-10days.	Transmit mechanically and by aphid vector *i.e. Myzus persicae* and *Aphis craccivora* in non-persistant manner. Interseasonal carryover occurs through infected crops and weeds.
Zucchini yellow mosaic virus (ZYMV), the unenveloped filamentous to flexuous *Potyvirus* with a modal length of 750 nm x 11 nm, (+)ssRNA, TIP: 55-60 °C, LIV: 3-5 days. Virus first reported in Cucurbita pepo from Italy (Lisa *et al.*, 1981).	Transmit mechanically and by aphid vector *i.e. Aphis gossypii, A. craccivora, Lipaphis erysimi* and *Myzus persicae* in non-persistant manner. Interseasonal carryover occurs through infected crops and weeds.
Bean yellow mosaic virus (BYMV), the unenveloped filamentous to flexuous *Potyvirus* with a modal length of 750 nm x 12 nm, (+)ss RNA, DEP 1:800-1:1000, TIP 56-60°C, LIV 24-32hrs.	Transmit mechanically and by aphid vector *i.e. Myzus persicae* and *Aphis fabae* in non-persistant manner. Interseasonal carryover occurs through infected crops and weeds.

Symptoms produced by the virus on pumpkin

WMV 1: Mosaic mottle, starting from the base of the lamina; irregularly distributed chlorotic yellowish patches on the surface of the lamina; laminar tissues rugose. In acute infection lamina becomes crumpled, blistered, yellowish, leaving isolated green tissues around the veins and veinlets.

Mixed infection of WMV 2, ZYMV and BYMV: Greenish yellow mosaic mottling on the upper surface of the lamina; chlorotic tissues mostly located near the margin of the lamina and internal areas; infected leaves become slightly deformed. Most characteristic symptom is thickening of vein and veinlets on the under surface of the lamina and formation of cup shaped enations; petioles becomes stiff and slightly curved.

Table 7. Viral diseases of ridge gourd in Eastern India

Particle characteristics and physical properties of viruses	Transmission and interseasonal carryover
Watermelon mosaic virus 1 (WMV 1), the unenveloped filamentous to flexuous *Potyvirus* with a modal length of 760-800 nm x 12 nm, (+)ss RNA, DEP 1:10000-1:30000, TIP 55-60°C, LIV 9-10days.	Transmit mechanically and by aphid vector *i.e. Myzus persicae* and *Aphis craccivora* etc. in non-persistant manner. Interseasonal carryover occurs through infected crops and weeds.
Zucchini yellow mosaic virus (ZYMV), the unenveloped filamentous to flexuous *Potyvirus* with a modal length of 750 nm x 11 nm, (+)ssRNA.	Transmit mechanically and by aphid vector *i.e. Myzus persicae* and *Aphis fabae* in non-persistant manner. Interseasonal carryover occurs through infected crops and weeds.
Cucumber green mottle mosaic virus (CGMMV), the unenveloped rod shaped *Tobamovirus* with a modal length of 364 nm x 18 nm, (+)ssRNA, TIP 80-90°C, LIV 1year.	Transmit mechanically and by contact; no vector known. Interseasonal carryover occurs through infected crops, weeds and contaminated seeds.

Symptoms produced by the virus on ridge gourd

WMV 1: In general, the lamina shows slight distortion, pale yellowing and irregularly distributed deep green blister. While in some cases of infection lamina becomes pale yellowish, with irregularly distributed deep green spots; vein and veinlets prominent.

ZYMV: Mosaic mottle, yellowish chlorotic spots on the surface of the lamina; veins and veinlets prominent and greenish.

CGMMV: Diffused yellowish chlorotic patches in the interveinal regions.

Table 8. Viral diseases of pointed gourd in Eastern India

Particle characteristics and physical properties of viruses	Transmission and interseasonal carryover
Cucumber mosaic virus (CMV), the isometric *Cucumovirus,* 30nm in diameter, (+)ssRNA, DEP 1:10000, TIP 60-70°C, LIV 3-4days.	Transmit mechanically and by aphid vector *i.e. Myzus persicae* and *Aphis gossypii* in non-persistant manner. Interseasonal carryover occurs through infected crops, weeds and contaminated seeds.
Cucumber green mottle mosaic virus (CGMMV), the unenveloped rod shaped *Tobamovirus* with a modal length of 364 nm x 18 nm, (+)ssRNA, TIP 80-90°C, LIV 1year.	Transmit mechanically and by contact; no vector known. Interseasonal carryover occurs through infected crops, weeds and contaminated seeds.

Symptoms produced by the virus on pointed gourd

CMV: Lamina shows mosaic mottling, irregularly distributed yellowish patches and green veinbanding.

CGMMV: Lamina shows irregularly distributed chlorotic spots.

Table 9. Viral diseases of cucumber in Eastern India

Particle characteristics and physical properties of viruses	Transmission and interseasonal carryover
Cucumber mosaic virus (CMV), the isometric *Cucumovirus,* 30nm in diameter, (+)ssRNA, DEP 1:10000, TIP 60-70°C, LIV 3-4days.	Transmit mechanically and by aphid vector *i.e. Myzus persicae, Aphis fabae* and *Macrosiphum euphorbiae* in non-persistant manner. Interseasonal carryover of the virus occurs through infected crops, weeds and contaminated seeds.
Potato virus Y (PVY), the unenveloped filamentous *Potyvirus* with a modal length of 750 nm x 12 nm, (+)ssRNA, DEP 1:100-1:1000, TIP 52-55°C, LIV 24-48hrs.	

Symptoms produced by the virus on cucumber

Mixed infection of CMV and PVY: Lamina shows mosaic with irregularly distributed chlorotic tissues with or without sharp margins.

Table 10. Viral diseases of bitter gourd in Eastern India

Particle characteristics and physical properties of viruses	Transmission and interseasonal carryover
Zucchini yellow mosaic virus (ZYMV), the unenveloped filamentous to flexuous *Potyvirus*, (+)ssRNA.	Transmit mechanically and by aphid vector *i.e. Aphis gossypii, Aphis craccivora, Lipaphis erysimi,* and *Myzus persicae* in non-persistant manner. Interseasonal carryover occurs through infected crops and weeds.
Potato virus Y (PVY), the unenveloped filamentous to flexuous *Potyvirus*, (+) ssRNA, DEP 1:100-1:1000, TIP 52-55°C, LIV 24-48hrs.	Transmit mechanically and by aphid vector *i.e. Myzus persicae, Aphis fabae* and *Macrosiphum euphorbiae* in non-persistant manner. Interseasonal carryover occurs through weeds and infected potato tuber.

Symptoms produced by the virus on ridge gourd

Mixed infection of ZYMV and PVY: Irregular yellowish chlorotic spots, gradually coalesce together to form more or less sharp yellowish areas particularly around the tips of the lamina becomes rough, thick and leathery with upward cupping; deep green blisters are found around the vein.

Table 11. Viral diseases of ivy gourd in Eastern India

Particle characteristics and physical properties of viruses	Transmission and interseasonal carryover
Watermelon mosaic virus 1 (WMV 1), the unenveloped filamentous to flexuous *Potyvirus*, (+)ssRNA, DEP 1:10000-1:30000, TIP 55-60°C, LIV 9-10days..	Transmit mechanically and by aphid vector *i.e. Myzus persicae, Aphis craccivora* and *Macrosiphum euphorbiae* in non-persistant manner. Interseasonal carryover occurs through infected crops where they harbour and transmitted to susceptible crops by the insect vectors/ mechanical contact.
Zucchini yellow mosaic virus (ZYMV), the unenveloped filamentous to flexuous *Potyvirus*, (+)ssRNA.	Transmit mechanically and by aphid vector *i.e. Aphis gossypii, Aphis craccivora, Lipaphis erysimi,* and *Myzus persicae* in non-persistant manner. Interseasonal carryover of the virus through infected crops and weeds. Interseasonal carryover occurs through infected crops

	where they harbour and transmitted to susceptible crops by the insect vectors/ mechanical contact.
Cucumber green mottle mosaic virus (CGMMV), the unenveloped rod shaped *Tobamovirus* with a modal length of 364 nm x 18 nm, (+)ssRNA, TIP 80-90°C, LIV 1year.	Transmit mechanically, no vector known. Interseasonal carryover occurs through infected crops, and acts as resurvour host of the virus for other cucurbits.

Symptoms produced by the virus on ridge gourd

Mixed infection of WMV 1 and ZYMV: Lamina shows diffuse mottling, mild deep green blisters; marginal lobes become deep.

CGMMV: Lamina shows irregularly distributed sharply marginal yellowish cholrotic spots.

Management of viral diseases

Continuous cultivation of cucurbits, monocropping and their intercropping provide ideal condition for the perpetuation, inoculum buildup and spread of the virus diseases.

The most important cucurbit virus in India is PRSV-W. It is efficiently transmitted by aphids in a nonpersistent manner and infects almost all the cucurbits grown in India. Separation of PRSV-W and WMV-2 is rather difficult. Therefore, for management purposes they may be considered as one virus.

Another important cucurbit virus is CGMMV, which again infects all the cucurbits. it is transmitted easily by contact. The epidemics of whitefly-transmitted geminiviruses in recent years are a pointer to the possibility of increase in diseases caused by geminiviruses necessitating great attention to their management.

Two other cucurbits infecting viruses which transmitted mechanically are muskmelon vein necrosis virus and squash mosaic virus.

Generalized management practices that can be used for minimizing the losses caused by cucurbit viruses are discussed below.

i) Disease free seeds/planting materials

In India, seed transmission has been found only for the CMV in different cucurbits. Whereas, in Japan CGMMV has been shown to be transmitted through seeds (Komuro *et al.*, 1971). Seed transmission of a geminivirus causing leaf distortion in bitter gourd has also been reported. So, disease free seeds from healthy plants are essential to avoid seed borne inoculum. To eliminate CMV infection in *Cucurbita pepo,* the virus can be inactivated by treatments with hot air (70^{0}C for two days) or hot water (55^0 C for 60 min) (Sharma and Chohan, 1973).

ii) Cultural practices

Field should be kept weed free. CGMMV can be transmitted through water (Vani and Varma, 1988). It is, therefore, essential to use uncontaminated water for irrigation. CGMMV also spread by contact and contaminated tools. Proper hygiene can be maintained to minimize spread of the virus. Avoidance of over lapped cropping will help reduce the incidence of viruses transmitted by aphids or white flies. It may also be possible to devise protection methods which minimise aphid transmission of the virus during the early period of growth (Thomas, 1980). Use of mulches (Vani *et al.,* 1989) can be reduced viral disease incidence and helps in improving the yields. Silver reflective mulches may reduce aphids' infestation (Kennelly, 2012). The light reflected from the mulch surfaces (aluminium foil or 'Panda' film) can prevent winged aphids from landing on the plants. If infection could be prevented or reduced until the plants started to crop, the loss in production could be minimised, as shown for WMV-2 on pumpkin and cucumber crops in New Zealand by Thomas (1971).

Rogueing and destruction of infected plants from the field is important to manage the disease. All the infected plants and perennial weed hosts should be eradicated from the field. The field and its surroundings should be kept free from the hosts including weeds. Control of WMV -1 is difficult since the presence of wild pumpkin and balsam apple facilitates a carry-over of the virus from one season to the next. Incidence of the virus can be reduced by destroying these weed hosts (Thomas, 1980). New planting near older cucurbits fields should be avoided. Proper sanitary measures should be maintained *i.e.* disease inoculum should be destroyed by ploughing down of old cucurbit crops.

iii) Prevention of vector transmission

The insect vector can be managed by the use of insecticides. Application of systemic insecticides to crops is unlikely to reduce the incidence of WMV-l 'because aphids may not be destroyed quickly enough to prevent their transmitting the virus in the non-persistent manner when migrating from infected weeds (Thomas, 1980). Target specific and eco-friendly insecticide should be used, as the insecticides are one of the main causes of pollution. It discourages pollinators and natural enemies in crop fields, and also disturbed soil microbiota. Application of oils (Krishi oil @ 2%) can be advantageous in reducing virus transmission by aphids in case of WMV-2 in *Cucurbita pepo* (Raychaudhuri and Varma, 1983). Poison bait (composition for preparation of 2 litre poison bait: Malathion 50% EC 20ml + molasses 200g + water 2 litre) can also be used to prevent insect vector effectively (Bhattacharya *et al.*, 2006). The disease can be controlled by spraying of Phosphamidon 40% SL @ 0.15% that effectively lower down the insect vector populations in field (Gour *et al.*, 2008).

iv) Inactivation of viral particle

The bottle gourd isolate of the CMV is completely inactivated by crystal violet, sodium lauryl sulphate and thiouracil, when mixed with the inoculum. Treatment of systemic hosts with sodium lauryl sulphate and thiouracil delays appearance of symptoms and the treated plants contain less virus than the untreated control (Rao *et al.*, 1976, Varma and Giri, 1998).

v) Host resistance

The majority of cucurbits possess huge genetic diversity. However, very limited efforts have been made to identify sources of resistance and development of resistant varieties. There is an urgent need to improve germplasm collection of various cucurbits so that resistant varieties are developed by traditional breeding (Varma and Giri, 1998).

2.5.4 Diseases Caused by Phytoplasma

2.5.4.1 Phyllodys

Phyllody is one of the most important diseases of cucurbits. Phyllody of bitter gourd and bottle gourd is common around Jaipur, Rajasthan (Misra and Gupta, 1988). Fluorescent microscopy indicated association of *Phytoplasma* (previously known as MLO) with the disease. Chow-chow is also infected by a graft transmissible disease and common in Kalimpong district of West Bengal (Ahlawat and Kulshreshtha, 1977), but involvement of insect vector for transmission of the disease and etiology of the pathogen have not been determined (Varma and Giri, 1998).

The disease is characterized by shortening of internodes and various parts of the flower turn into leafy structures. Some plants bear cluster of phyllody flowers. The leaves of the infected plants are distorted. Infected plants remain stunted and usually do not bear fruits. The disease infects almost all cucurbit vegetables.

Phytoplasmas are pleomorphic, small, rounded, large globular and branch filamentous, one or two to several microns in length and their average diameter is 0.3 to 0.8 microns.

2.5.4.2 Witches broom

Witches' broom of bitter gourd was first noticed around Bangalore (Singh, 1985, Singh, 1992). The incidence of the disease varies from 20 to 60 per cent (Varma and Giri, 1998).

The disease is characterized by chlorotic little leaves, virescence, phyllody, and proliferation of axillary buds. The fruits of late infected plants are small, thin, and cylindrical. The cultivar Coimbatore Long is very susceptible (Varma and Giri, 1998). Plants infected at early stages show total loss of yield and even in late infection, the yield losses are 30-50 per cent (Singh, 1985).

The causal agent is suspected to be *Phytoplasma*. It transmitted through graft, but not by mechanical inoculation, whitefly, or aphids (Varma and Giri, 1998). More research needed to detail about the transmission of the disease.

Management of the diseases caused by phytoplasma

a) All the diseased plants must be rouged out from the field.

b) Application of Carbofuran 3%G @ 30 kg/ha or Cartap hydrochloride 4% G @ 25kg/ha at the time of sowing the seed is advisable to manage the vector.

c) Spraying of systemic insecticides like Acetamiprid 20% SP (0.05%) at 10 days interval.

d) Disease symptoms can be suppressed by five sprays of 500 ppm oxytetracycline hydrochloride solution at seven days intervals (Varma and Giri, 1998).

2.5.5 Post Harvest Diseases

It is well established that the cucurbits are grouped under perishables suffering from several fungal and bacterial diseases under field conditions show fast deterioration after harvest. Generally, the quality of perishable vegetables deteriorates within 10 days of harvest, which consequently accelerate by the attack of microbes. Damage like blemishes or bruishes occurs during havest, handling or transportion of vegetables make them more prone to microbial invasion (Roy and Singh, 2003). An account of

8-15 per cent post harvest loss in some cucurbits was recorded by Choudhary (1986). Post harvest deterioration of parishables are usually associated with various symptoms *viz.* soft rot, dry rot, pink rot, blue green mould rot, phoma rot, anthracnose, bacterial rot, fusarium rot and charcoal rot etc. The microorganisms which are responsible for spoilage of vegetables lowers the nutritive value in declining the amount of sugar during host-parasite interactions is probably due to breakdown of carbohydrates by the fungal enzymes, increases in the rate of respiration in the infected host tissue and utilization of host carbohydrates by the fungi for various metabolic activities. Proteins are also used by microorganisms to meet up their nutritional requirements for better growth and development. Besides, fungal infections lead to reduction in the concentration of ascorbic acid and somewhat minerals that actually bring about changes in their quantity and quality (Roy and Singh, 2003). Some important post harvest diseases of cucurbit vegetables are discussed below.

2.5.5.1 White cottony rot

The disease is mainly caused by *Fusarium equseti, F. oxysporum, F. solani* (Mart.) Appel & Wr. on pointed gourd and little gourd. These pathogens were reported from West Bengal and Bihar respectively (Chattopadhyay and Mustafee, 1967, Mandal and Dasgupta, 1981, Prasad and Podder, 1977). Other *Fusarium* spp. was also recorded from different parts of the India. The disease of little gourd due to *F. semitectum* Berk. and Ravenel has been earliar recorded from Karnataka (Hiremath and Govindu, 1973). *Fusarium equseti* (Corda) Sacc. is more prevalent species.

The disease is quite common all over the season but causes extensive damage during June-July and October-December. The rot starts from anywhere on the fruit surface but mostly from the poles as small water soaked spots. Rotted area becomes to some extent depressed and measures about 2-5 cm in diameter. Infected area is soft, dull and yellow and covered by white to peach, floccose growth which sometimes invades the hollow center of the fruits as well. Rotted fruits emit a peculiar offensive smell (Chattopadhyay and Mustafee, 1967).

It is one of the severe diseases of tinda and has been recorded from Rajasthan (Mathur and Mathur, 1958), Andhra Pradesh (Rao, 1966) and Madhya Pradesh (Chaurasia, 1980). It is seed borne in nature (Suryanrayan *et al.*, 1963). The disease appears on the rind as circular brown patches that spread rapidly. Afterwords, profuse white mycelial growth develop and cover the entire fruit surface. The whole fruit becomes soft and pulpy within 6-7 days. Internal rotting is common in injured fruits (Dasgupta and Mandal, 1989).

White or pink rot, a common post harvest disease of longmelon, caused by *F. semitectum, F. oxysporum* and *F. solani* have been recorded from Haryana and West Bengal (Mandal and Dasgupta, 1981, Dasgupta and Mandal, 1989). The disease occurs during April-June on fruit bores by insect larvae, or otherwise bruised. The rot starts around the injury as water soaked lesions progressing both dry and humid conditions. Afterward, a pinkish or cottony white mycelia mass appears on the fruit surface as well as within the hollow centre while the skin remains intact. Emittance of an offensive smell can be recorded only with a secondary infection due to bacteria. Larvae bores provide the pathogen an easy access and thereby a multisite infection can develop in nature. High relative humidity favours the disease (Mandal and Dasgupta, 1981, Dasgupta and Mandal, 1988). Pink cottony rot of cucumber caused by *F. moniliforme* was also noticed during winter (Mandal and Dasgupta, 1981). Infection starts from the weak pedicel end or around an injury. The rotted area is slightly depressed, progressing slowly on the surface with floccose pinkish growth covering the fruit surface either partially or fully, and finally transforming the fruit into a soft watery mass (Dasgupta and Mandal, 1989).

White cottony rot or spongy rot of pumpkin due to *Fusarium oxysporum* has been reported from Himachal Pradesh (Gangopadhyay and Sharma, 1976) and due to *F. solani* from West Bengal (Mandal and Dasgupta, 1980). *F. solani* f.sp. *cucurbitae* race 2 has become serious in the USA (Hall *et al.,* 1981). The disease observed on stored ripe fruits during June-July with a characteristics oak brown soft area around the injured site. The flesh of the infected fruits becomes combed brown. Infection progress radially with distortion of the fruit skin and shrinkage of the flesh, which causes cylindrical hole of about 2.5 to 5.0 cm diameter. Scanty mycelia growth can be observed on this half-rotten fruits (Dasgupta and Mandal, 1989).

Scanty cottony white rot disease is.common in *phuti* (*Cucumis melo* var. *phuti*). The rotting originates from the distal end of the fruit during April and May in the form of browning, depressed blemish covering measuring of about 5-6cm in diameter. Scanty mycelia mass can be seen from the commencement on the blemish over a rather shallow rotting but deep softening of the inner tissue. The disease is more pronounced on injured fruits than uninjured one and rate of spread is very slow (Dasgupta and Mandal, 1989). Taste of the infected fruit may vary from normal to extreamly bitter depending on the species of the pathogen involved (Ryall and Lipton, 1972). The disease caused by two fungal pathogens i.e. *Fusarium equiseti* (mycelium with red, blue or violet tinge) and *F. oxysporum* (pink myceliam and spore mass). Both the species cause similar decay symptoms. The pathogen disseminates through contact (Dasgupta and Mandal, 1989).

Cottony white rot of bitter gourd is quite common in India. The disease caused by different species of *Fusarium* was reported from Maharashtra, Karnataka, West Bengal and Uttar Pradesh (Rao, 1966, Hiremath and Govindu, 1973, Chattopadhyay and Mustafee, 1967, Dasgupta and Mandal, 1989, Tandon and Varma, 1964). *Fusarium* sp. *F. equiseti, F. oxysporum, F. solani* sp. *cucurbitarum* have been reported to be infected the crop. *F. equiseti* is the prevalent one (Dasgupta and Mandal, 1989). Both young and matured bitter gourd fruits are infected by the disease throughout the year. Maximum damaged of the crop occurs during summer month. The disease initiates from the distal end that characterized by a little water soaked circular to irregular rotten patch with withered ribs and skin. Slightly depressed rotted zone encircled by a dark zonation of about 1cm in diameter, which finally measuring to about 5-6cm in diameter is the important diagnostic symptoms. Under dry conditions the fruits become hard, but under favourable conditions internal tissue disintegrates with a brownish shade. The reddish slimy host tissue on the seed surface become disintegrated and separated. Seeds surface is invaded by a fungal spores mass. Afterwords, whitish to dull white scanty of fluffy cottony mycelium develops on the fruit surface in moist atmosphere. Secondary infection by the bacteria cause quick decay and leak out of juice from infected fruit without any unpleasant odour. Injured fruits are easily infected by the pathogen (Dasgupta and Mandal, 1989).

Fluffy white rot of *kakrol* or spiny gourd (*Memordica dioica* Roxb. Ex Willd.) caused by *F. semitectum* and other *Fusarium* spp. has been reported from Madhya Pradesh and West Bengal (Gupta *et al.*, 1982, Mandal and Dasgupta, 1981). Injured fruits are suffered mainly that characterized by water soaked spots with softening of the tissue. In advanced stages of infection fluffy cottony mycelial growth develops over the fruit (Dasgupta and Mandal, 1989).

Cottony rot of bottle gourd has been reported from different parts of India. The inciting agent is *Fusarium* spp. The reported species from Bihar are *F. concolor, F. equiseti* and *F. semitectum* (Roy, 1973), from Punjab are *F. semitectum, F. dematium* and *F. solani* (Sohi, 1984, Singh and Chohan, 1980), and from Haryana is *F. equiseti* (Sumbali and Mehrotra, 1982). The characteristic symptom of the disease is discrete or coalescent watersoaked lesions that appear on the rind of mature and ripe fruits. Slightly depressed pinkish or white mycelial growth develops over the surface (Dasgupta and Mandal, 1989).

Different species of *Fusarium* viz. *F. oxysporum, F. solani, F. equiseti* and *F. moniliforme* etc. have been reported as white cottony rot causing fungi from West Bengal, Bihar and other parts of India (Chattopadhyay and Mustafee, 1967, Mandal and Dasgupta, 1981, Prasad and Podder, 1977. Among them, *F. solani* and *F. oxysporum* are more common and often co-habitating (Dasgupta and Mandal, 1989). The disease is quite common all over the season but causes extensive damage during June-July. The rots starts mostly from the poles, but may appear anywhere on the fruit surface. The rotted area depressed slightly that measures about 2-5cm in diameter. Infected area become soft, dull and yellow, and covered with whitish to peach, floccose growth that occasionally invades the hollow centre of the fruit too. Rotted fruits emit a characteristic smell (Dasgupta and Mandal, 1989).

2.5.5.2 Soft rot

The disease was recorded on cantaloupe in the USA and the pathogen was identified as *Rhizopus* spp. (Ramsey and Smith, 1961).

Dasgupta and Mandal (1989) described the disease in their book, Postharvest Pathology of Parishables. According to them, over ripe fruits of *phuti* are more affected than semi ripe ones in the markets in West Bengal during March to May. Rotting becomes prominent as a water soaked irregular depression of about 5-6 cm in diameter on the surface. Natural maize yellow colour of the fruit tarnishes into cadmium yellow in due course of decay. Inner edible pulp becomes to some extent softened and reddened. The disease is frequently first discovered at this stage when a finger pressed suddenly plunges into the soft flesh. In severe cases brown and thin mycelia mass invades the core of the fruit wherein the seeds are embedded. The rot progress very fast and putrefies the fruit within 3-5 days.

R. stolonifer cause soft watery rot on bitter gourd that occur during rains and was recorded from Karnataka and West Bengal (Hiremath and Govindu, 1973, Mandal and Dasgupta, 1981, 1982d). The disease begins as a small, water soaked area around an injury, progress rapidly become yellowish-brown and being covered by a stringy coarse fungal mass. In advanced case, the surface becomes very soft, and because of tissue disintegration watery rot sets in. Mycelial mass invades the seed coat. Infection develops in pockets consisting of a few fruits in the bucket if the primary foci remain unnoticed. The whole fruit is putrefied into a soft mass within a week under congenial conditions. Affected fruit emits a bad odour. Touch with the infected fruits is possibly responsible for fast dissemination of the disease (Dasgupta and Mandal, 1989).

Soft rot of longmelon caused by *R. rhizopodiformis* was reported from West Bengal (Mandal and Dasgupta, 1981). The disease is quite common in the markets during March, mostly on bruised and insect-injured fruits. At first, slightly discoloured lesion appears on the fruits, which becomes circular to oval in shape. The advancing margin of the lesion is light greenish-yellow and water soaked. The surface of the affected fruit becomes soft with sparse mycelial growth. Whereas, under humid conditions dense stringy mycelial mass partially covers the fruit surface in a short time and gets intermingled with the seed. A bad odour may emit from the infected fruit due to secondary bacterial infection (Dasgupta and Mandal, 1989).

Soft rot of ridge gourd has been described from Punjab and West Bengal (Singh, 1974; 1979, Mandal and Dasgupta, 1981). The causal organism of the disease is *Mucor hiemalis* f.sp. *luteus, M. circinelloides*). This is not a common problem. Three to four days old fruits damaged by insects or physical means are affected severely during July-September. *Mucor hiemalis* also caused fuzzy yellowish rot of bitter gourd mostly during October to November (Dasgupta and Mandal, 1989). Water soaked patches appear around the injured part of the fruits followed by appearance of a yellowish fuzzy mycelial growth. The rotted portion becomes slightly shriveled, soft, dull in colour, and emit a bad odour. The rotting due to the *Mucor* spp. is slower than the cottony leak. The disease favoured by high humidity. In some cases, the disease associated with *Fusarium* spp. (Dasgupta and Mandal, 1989).

Soft rotting due to *Fusarium* spp. reported earlier on snake gourd and ridge gourd (Rao, 1966). Mandal and Dasgupta (1981) noticed the disease during July-August in West Bengal. The disease occurs rarely in snake gourd. Injured fruits are affected mainly, and a small water soaked specks appears around the injury that progress rapidly both radially and into the edible tissue. The advancing margin is lavender green whereas the inner central has a small red zonation, about 2-3mm in diameter. Afterwards, scanty, white or pinkish, floccose mycelium develops on the infected fruit surface. Cracking of fruits at the infected site can also be noticed. The fungus sporulates in the hollow centre of the fruit (Dasgupta and Mandal, 1989).

They also mentioned the soft rot of muskmelon caused by *Pythium butleri* Subramaniam. Singh and Chouhan (1977) reported that the disease was recorded from Punjab. During transportation and marketing the injured fruits suffer seriously. The disease primarily appears as minute dots, which gradually enlarge to a water soaked area of about 2-4 cm diameter on the upper end of the fruit. While the colour turns from faded green to brown, the whole fruit becomes soft within 2-3 days and is covered by fluffy cottony white mycelium. Possibly the fungus is carried from the field.

Soft rot of little gourd caused by *Choanephora* sp. is a minor disease noted only during November particularly on the produce delayed in transit (Mandal and Dasgupta, 1981). The infected fruits primarily show watersoaked spots of champagne colour. Rotting spreads quickly with dense mycelial overgrowth, extensive internal ramification and a characteristics odour.

2.5.5.3 Calyx end rot

Rao and Subramaniam (1975) were observed the disease on muskmelon in Maharashtra during March. The disease initialy appears as irregular, slightly depressed, blackish, rotted lesions at the calyx end, which spread rapidly and the fungal growth covers the lesion surface. In early stage, the epicarp becomes yellow and finally turns brownish starting from the centre. The flesh becomes soft and loose, and emits fermentative fruity odour. The causal agent of the disease is a fungus, *Thielaviopsis paradoxa* (Dasgupta and Mandal, 1989).

2.5.5.4 Cottony leak

It is a common and most troublesome disease of pointed gourd, cucumber, snapmelon and tinda in India. The disease continues to be serious during transit, storage and marketing. It is aggravated by bruises and many weak parasites become a part of the causal complex. The disease is common in occurrence under humid conditions during rains. *Pythium aphanidermatum, P. butleri* Subramaniam *and P. cucurbitacearum* S Takim. have earlier been reported in India especially from Bihar and West Bengal Mehrotra, 1954, Vasudeva, 1960, Chaudhuri, 1975, Mandal and Dasgupta, 1981). Cottony leak of snapmelon has been recorded from Tamil Nadu and ridgegourd and bittergourd from Maharashtra and West Bengal (Rao, 1966, Mandal and Dasgupta, 1981). In pointed gourd, rotting generally starts on infected fruits during March-September, being more serious on the physiologically weakened fruits while the cottony leak of cucumber is prevalent from April to August being severe during July to August (Dasgupta and Mandal, 1989). The disease is quite common in ridge gourd during May to September but is severe during May to June whereas that was noticed on bitter gourd during July to August (Dasgupta and Mandal, 1989). Infection appears suddenly in pockets of eight to ten fruits within bamboo basket in the form of water soaked light green areas anywhere of the fruit surface, but certainly to be covered overnight by fluffy cottony white mycelial mass with contaminant growth inside as soft rot sets in (Khatua and Saha, 2004).

The disease can be noticed in high proportion during August-September on the long-transported produce of tinda. It manifests itself as a small watersoaked lesion on any part of the fruit surface, which progress rapidly but irregularly. Normal green colour of the fruit gradually fades out and the lesion becomes slightly brownish with a cottony fluffy overgrowth which covers the whole fruit within 36 hours (Dasgupta and Mandal, 1989).

The infection is possibly carried over from the field. High humidity favours quick rotting. Mature fruits are more susceptible than young ones (Dasgupta and Mandal, 1989). Fruit decay in storage and transit can be prevented by treatment of imazalil (2000ppm) along with waxing of fruits. Aharoni *et al.* (1993) suggested treating fruits with hinokitiol, a volatile oil extracted from the Japanese Hika tree (*Thujopsis dolabrata*).

2.5.5.5 Waxy rot or Sour rot

This minor disease is caused by *Geotrichum candidum* Link recorded from Maharashtra (Rao, 1966) and West Bengal (Mandal, 1981). Waxy rot of tinda suffer most during July-August whereas, in pointed gourd and pumpkin during June-October (Mandal, 1981, Mandal and Dasgupta, 1981). The disease is almost invariably associated with semi-ripe, physiologically weak or bruised fruits. Sour rot is also noticed only on a day-old left over stock of injured ridgegourd fruits in the market or kitchen store. Infected fruits lose its lusture (Mandal and Dasgupta, 1981). Infection develops in patches as white powdery growth. Under humid conditions it become dirty and looks like a bacterial colony on the fruit surface. In this condition the infection progress fast and the infected fruit subsequently becomes soft. Such fruits ooze out juice that carry spore laden mass and emit strong offensive odour (Khatua and Saha, 2004). *Fusarium* sp., when associated with this disease, produces scanty mycelia growth on the fruit surface. Both the organisms are individually pathogenic on tinda (Mandal and Dasgupta, 1981). *Bacillus* spp. and yeasts are also associated with the sour rot of pumpkin (Mandal and Dasgupta, 1983c).

2.5.5.6 Anthracnose

The infected cucurbit fruits show remarkable discolouration followed by appearance of water soaked patches, in the initial stages. The infected regions become yellowish white, which gradually turned black. A large number of small pinheads like structures i.e. acervuli appear on affected areas of the fruits. The spot turns pinkish white due to spore production in acervuli. The infected fruits finally fall off (Sahu Kritagyan and Singh, 1980). The causal organism was isolated, identified and confirmed as

Colletotrichum state of *Glomerella cingulata* (stoneman) Spauld & H. Shrenk. (Sahu Kritagyan and Singh, 1980).

Anthracnose of snapmelon reported only from Karnataka caused by *Colletotrichum capsici* (Syd.) E. J. Butler & Bisbey and *C. lagenarium* (Pass.) Ell. et Halst. is an important disease produce circular to semi-circular water-soaked lesions measuring up to 2.5-9mm in diameter. These lesions may coalesce and became light yellowish-brown in colour and sunken (Dasgupta and Mandal, 1988).

Postharvest anthracnose of bittergourd has been recorded from Karnataka (Prakash and Singh, 1977). The disease is caused by *Colletotrichum lagenarium*. The infected fruits show abundant small, circular, watersoaked, sunken or depressed area over the surface. The spots are straw to grayish brown in colour and merge into large elongate patches on the fruit. Afterward, the fruits crack open longitudinally, which comprise the most characteristic symptoms of the disease. Pink spore masses may develop over the surface but inner tissues remain free from infection (Dasgupta and Mandal, 1989).

Anthracnose of snake gourd was recorded in Uttar Pradesh (Singh, 1973) as stem end rot that later renamed. Grayish, oblong to ovoid or irregular slightly sunken lesions appear on fruit surface, which frequently coalesce and spread longitudinally towards the distal end. The affected area turns light pink-green to yellow and finally become dark brown to black. Such fruits become soft, water-soaked and rotten totally. Under humid conditions, the acervuli may be developed on the rotted tissue (Mandal and Dasgupta, 1988).

Anthracnose of bottlegourd incited by *Colletotrichum lagenarium* is a devastating disease, occurs world-wide both in field as well as in market (Madan and Grover, 1977, Bilgrami *et al.*, 1981). The disease appears as water-soaked, circular to irregular, oval to oblong lesions that become brown to black and scabby. The lesions are coalesceing. Under high humid conditions, pinkish spore mass may be appeared on old scabby lesions (Mandal and Dasgupta, 1988).

To avoid spoilage of fruits care should be taken to remove infected and injured fruits before transit and arrangement is to be taken to keep the fruit surface dry (Khatua and Saha, 2004).

2.5.5.7 Brown rot

The disease is noticed on over-ripe fruits of *Phuti* (*Cucumis melo* var. *phuti*) in markets during May and June. The rotting is brown and progress fast radially and downwards. Later, the inner core is blackened and filled with blackish mycelium. Almost immediately, a quite fungus develops and covers the entire ripe fruit producing a mosaic effect. Severely infected fruits emit offensive smell. The fungal pathogen responsible for the disease is *Botryodiplodia theobromae*. Some other pathogens, which are associated with the disease, are *Aspergillus niger, A. flavus, Alternaria* spp., *Fusarium equiseti* and *F. oxysporum* (Dasgupta and Mandal, 1988).

Brown rot of snakegourd, caused by *Alternaria tenuissima* has been reported from Uttar Pradesh (Singh, 1974). Infection starts at the stem end as roughly circular grey spots of 0.3 to 0.4 cm diameter, which merge, become yellow to pink and finally turn brown. The tissues below the lesion may be dried up and hanged out from the healthy portion (Mandal and Dasgupta, 1988).

Brown spot due to *Alternaria alternata* develop on injured or uninjured bottlegourd as small circular to oval, dark coloured spots of 1.5 to 2.5 cm in diameter with depressed centre and raised margin that may enlarge up to 6 cm (Dasgupta and Mandal, 1988).

Brown rot of little gourd caused by *Macrophomina phaseolina* is a minor disease occurring in June-July on green to semi ripe fruit (Mandal and Dasgupta, 1981, Mandal, 1981). The diseased fruit shows a sunken circular to oblong spot of approximately 0.5 cm diameter at the distal end. Infection is skin-deep but enlarges superficially with scanty whitish mycelial mass within the sunken area. Brown micro sclerotia appear later with the area becoming blackish brown. The skin remains intact but the fruit either becomes hard or soft depending on the ambient moisture. Fruits near the ground level possibly get contaminated in the field itself (Dasgupta and Mandal, 1989).

2.5.5.8 Green mould rot

The fruits of watermelon injured during road transportation are almost regularly affected by the disease during May-June in West Bengal (Mandal, 1981). The cracked fruits are mainly affected. The surface of the crack zone covers with greenish fungal growth. The red colour pulp of the fruit becomes fade. The rotting caused by the fungus, *Aspergillus fumigates* Fresenius is very slow and limited, both in natural infection as well as artificial inoculation. The causal agent is expected as a week pathogen (Dasgupta and Mandal, 1989).

Aspergillus fumigates causes disease on matured and injured fruits of bottlegourd as water soaked area that soon covered by a fungal mass with conidia and conidiophores (Dasgupta and Mandal, 1989).

The longmelon also infected with green mould causing rot during summer season. The fruits injured by larvae or otherwise bruised are affected mainly (Mandal and Dasgupta, 1981). Infection starts as a soft area around an injury. The rot progress very fast and the conidial heads of the causal fungus (*Aspergillus foetidus* var. *acidus* Nakazawa et al., *A. fumigates*) appear on the injured surface. The rot either progress or arrested in high or low relative humidity, respectively (Dasgupta and Mandal, 1989).

Aspergillus niger van Tieghem causing disease on pumpkin called black mould rot. This is a minor disease that occurs on the cut pieces or on rat-browsed fruits of ripe pumpkin (Dasgupta and Mandal, 1989).

2.5.5.9 Stem end rot

Stem end rot of watermelon is caused by a fungal pathogen, *Botryodiplodia theobromae* Pat. The disease has earliar been recorded from the USA (Walker and Weber, 1931). This serious disease occurs in West Bengal on semi-ripe fruits during May-June on the marketed produce. The infection initiates at the stem end with watersoaked appearance, which become a Naples yellow rim and an extream ashy grey centre. Afterward the infected fruit rots very rapidly, becomes wrinkled and dried up, with ashy-grey mycelial mass over the surface. The inner tissues turn Vandyke brown, emit a fermentative odour and leak a watery juice. Infection may be carried from the field or consequently through contact (Dasgupta and Mandal, 1989).

The pathogen, *Botryodiplodia theobromae* also caused brown pedicel end rot of snapmelon during summer months (Mandal and Dasgupta, 1981, Mandal and Dasgupta, 1982). The infected fruit shows patches of small brown area at the pedicel end that coalesce to form large spots or lesion and progress with its brown coloured centre and yellowish watersoaked margin. Sparse ashy grey mycelial mass later turn brown, and grow outside and into the inner cavity entangling the seeds. The inner tissues become soft and watery. Such affected fruits emit characteristics fermentative smell. The juvenile tissues appear to be resistant (Mandal and Dasgupta, 1989).

2.5.5.10 Charcoal rot

This is a common and serious disease of snapmelon and ridgegourd caused by *Macrophomina phaseolina* known in Punjab (Jhooty and Singh, 1971, Singh and Chohan, 1972) and West Bengal (Mandal and Dasgupta, 1981). Infection initiates generally from the distal end of the fruit as water soaked area with varying shades of brown and later turns blackened. Abundant sclerotia develop as the rotting progress towards the pedicel end with shrinkage of the skin and tissue. Under moist conditions a skin-deep superficial blemish develops as depressed blackish growth associated with fast rotting. A dry spell can result in halting the disease (Dasgupta and Mandal, 1989).

2.5.5.11 Grey white rot

The disease of pumpkin is caused by *Sclerotium rolfsii*, noticed from April to August mainly on ripe and injured fruits (Mandal and Dasgupta, 1981). The infected fruits show a circular, slightly depressed area of about 0.6 to 1.5 cm in diameter, progress slowly but occasionally enter tissue instead of expanding radially. It becomes somewhat soft, watery and dull. The pathogen possibly finds its avenue through insect bites or other injuries (Mandal and Dasgupta, 1988). The disease develops as small localized patches with superficial creepy shining white mycelial overgrowth, which is shortly strewn with copious sclerotia (Dasgupta and Mandal, 1981).

Grey white rot of bittergourd, one of the worst postharvest diseases, recorded in Karnataka and West Bengal (Singh *et al.*, 1973, Mandal and Dasgupta, 1981). The disease occurs mainly on large sized bittergourd between April and July. Small, slightly depressed, watersoaked areas appear on the fruit surface around a deep injury, which become ashy yellow and soft finally turn grayish to blackish. The growth of the fungus is limited in dry weather condition but conspicuous chalky white stranded mycelial mass creeps all over the fruit surface without forming sclerotia in high humid conditions. Infection may be initiated from the field that spreads by contact and develops in pockets. A bad odour emits from the infected fruit if secondary bacterial invasion takes place (Mandal and Dasgupta, 1989).

2.5.5.12 Dirty grey rot

Dirty grey rot incited by *Rhizoctonia solani* occurs in ridgegourd during August to September and bittergourd during October to December (Dasgupta and Mandal, 1980, 1981). Infection on ridge gourd usually begin from the distal end showing a depressed area of about 4-5 cm in diameter with a slightly water soaked advancing margin. Rotting progress rapidly and contaminate neighbouring fruits. Water congestion may be noticed within the affected areas, which subsequently develop straw colour. Dirty coarse mycelial mass develops afterwards rarely being intermingled with sclerotia measuring 0.3 to 0.5 cm in diameter, the pathogen is perhaps carried incipiently from the field by those fruits touching the ground. On bittergourd, the rot initiates from anywhere on the surface of fruit as watersoaked patches of irregular area, progress fairly into dull blackish patches being covered by extrametrical, coarse, light yellowish mycelium generally without sclerotia or any unpleasant odour. The infection may be carried from the field.

2.5.5.13 Black spot

The disease is recorded on both injured and uninjured matured bottlegourd as brown to black spots with irregular margin on the rind. Afterwords, the lesions are covered with a velvety black fungal growth accompanied by shallow rots. The causal agent of the disease if *Curvularia ovoidae* (Dasgupta and Mandal, 1989).

2.5.5.14 Physiological ripening / Yellowing

It is one of the important post harvest diseases of cucurbits, which causes deterioration of the harvested products in the marketing process in West Bengal. The disease is very much common in little gourd, bittergourd etc. In case of littlegourd ripening starts mainly from the pedicel and progresses very fast, consequently reducing the market price significantly, and the fruits become insipid while, in bittergourd and pointedgourd yellowing start mostly from the pedicel end until the entire skin becomes bright yellow. In the meantime, the inner mass turns red or yellow and the infected fruits may burst along the ridges from the distal end. Physiologoical yellowing may be due to hot dry weather is quite common in bittergourd during March-April however that is in pointed gourd during March-September with a severe form during April-June (Mandal, 1981a, 1981b, Dasgupta and Mandal, 1988).

Roy and Singh (2003) reviewed in detail about the post harvest diseases of cucurbits in Indian condition and prepared an exhaustive list as follows:

Table 12. List of fungi and bacteria causing post harvest diseases on cucurbit vegetables.

Host	Pathogen	Symptoms
Lagenaria siceraria	*Fusarium equiseti*	Soft rot
	F. semitectum	Dull yellowish water soaked patches
	F. scirpi	Soft rot
	Botryodiplodia theobromae	Dry rot
	Drechslera australiensis	Blackish-brown lesions
	Alternaria alternate	Circular blackish brown spots
Luffa cylindrica	*F. equiseti*	Soft rot
	F. scirpi	Soft rot
	A. alternate	Deep green irregular spot
	D. australiensis	Blackish brown water soaked lesions
	B. theobromae	Small black and soft lesions
	Myrothecium roridum	Dull-black water soaked irregular spots
	Helminthosporium spiciferum	Soft rot
	Curvularia lunata	Greenish white water soaked irregular lesions
	Pythium aphanidermatum	Water soaked areas with white cottony mycelia mat
	Aspergillus flavus	Soft rot
	Rhizopus stolonifer	Soft rot
	Alternaria alternate	Soft rot
	F. solani	Dry rot
Momordica charantia	*F. oxysporum*	Soft rot
	A. tenussima	Soft rot
	D. spicifer	Dull-black irregular patches
	A. alternate	Circular lesions, surrounded by water soaked areas
	C. lunata	Small dull-black regular lesions with greenish tinge at the centre.
	Aspergillus flavus	Greenish patches
	A. niger	Soft rot
	Rhizopus stolonifer	White mycelial growth
	Myrothecium roridum	Fruit rot

Trichosanthes dioica	*F. equiseti F. semitectum*	Soft rot
	F. moniliforme	Soft rot
	P. aphanidermatum	Soft rot
	C. lunata	Soft rot
	D. australiensis	Soft rot
	B. theobromae	Soft rot
	Colletotrichum gloeosporioides	Soft rot
	Alternaria alternate	Soft rot
	Aspergillus flavus	Soft rot
	A. niger	Soft rot
	Rhizopus stolonifer	Soft rot
	Rhizoctonia bataticola	Soft rot
	Cylindrocapon tankinense	Soft rot
Citrullus vulgaris var.*fistulous*	*Curvularia lunata*	Blackish brown round patches
	Alternaria alternate	Round to irregular dull brown erumpent patches
	F. scirpi	Circular to irregular pale brown patches
Cucumis melo var. *momordica*	*F. semitectum*	Dull yellowish water soaked patches
	Rhizopus solani	Fruit rot
	Myrothecium roridum	Fruit rot
	Pseudomonas lachrymans	Bacterial spot
Coccinia indica	*F. moniliforme*	Dull green oval to irregular water soaked areasas well as soft rot
	D. spicifera	Dull green to darty black small patches
	F. scirpi	Round irregular pale area
	Colletotrichum dematium	Soft rot
	A. alternate	Small regular brown coloured spot
	B. theobromae	Dry rot
	Aspergillus flavus	Green colonies
	A. niger	Soft rot
	R. stolonifer	Luxuriant mycelial growth
Cucurbita moschata	*Rhizoctonia solani*	Fruit rot
	F. oxysporum	Fruit rot

2.5.6 Management of Post Harvest Diseases

Management of post harvest diseases is very difficult. A large number of fungicides, fumigants and other pesticides are being extensively used to manage the post harvest diseases. Some of those fungicides are non-degradable and have residual effects. Many researchers tried to manage the diseases by different ways. But no one is effective alone. So, integrated management option is the best way. The following points should be remembered for effective management of the post harvest diseases of cucurbits:

a) To avoid spoilage of fruits care should be taken to remove infected and injured fruits before transit and arrangement is to be taken to keep the fruit surface dry (Khatua and Saha, 2004).

b) Though the inoculum of most of the post harvest diseases comes from field by contact with soil so scaffolded cultivation or use of mulching materials can be an option to avoid the diseases.

c) Adhering soil on the fruit surface should be removed.

d) Paper wraps alone or impregnated with fungistatic chemicals check the spread of pathogens associated with post harvest diseases. Fruits can individually be wrapped in heat shrinkable polyethylene film, which prevents the spread of disease and transmission of spores (Roy and singh, 2003).

e) *Fusarium* spp. infecting longmelon can be managed through pre- and post-inoculation treatment with bleaching powder and Potassium permanganate @ 0.5 to 1.0% (Dasgupta and Mandal, 1988).

f) Dithane M-45 (0.2%) revealed most effective for controlling fruit rot of bottle gourd caused by *Botryodiplodia theobromae* and *Fusarium semitectum* (Ojha, 1983).

g) Pre harvest spraing with captan (1.0%) can effectively manage the *Alternaria cucumeriana* (Dasgupta and Mandal, 1989).

h) *Pythium aphanidermatum* can be managed effectively by post-inoculation cotton swab treatment with JF 3937 (Wahab and Sharma, 1976), cotton swab and dipping treatment in multifungin, chloromycetin, aureofungin (50-200μg/ml) and actidione (10-12 μg/ml) in dimethyl formaldehyde for 3 days (Dasgupta and Mandal, 1989).

i) Some non-hazardous chemicals are extensively used to manage pre and post harvest diseases of perishable vegetables either by preventing sporulation or initiation of infection or by supressing the development of pathogens and latent infection in the host tissues (Roy and Singh, 2003).

j) Ojha (1983) and many other scientists have tested some plant extracts against diseases of bottle gourd, pointed gourd and other perishables vegetables. Besides, antimicrobial properties of a large number of angiospermic plants have been evaluated and reported by many researchers (Roy and Singh, 2003).

k) Attempts were made to mange the post harvest diseases of perishables using antagonistic microorganisms with quite success. Antagonistic microorganisms reduced inoculums load directly or indirectly from the plant or fruit surface and keep the products safe from post harvest rotting. Antagonists successful in controlling post harvest diseases included a number of bacteria *viz. Bacillus subtilis, B. licheniformis, B. stearothermobilius, B. megaterium, B. amyloliquefaciens, Pseudomonas corrugate* and the fungi *viz. Aureobasidium pullulans, Trichoderma viride and Penicillium funiculosum* (Roy and Singh, 2003).

l) To control *Pythium aphanidermatum* infecting cucumber, post inoculation treatment with undiluted autoclave culture filtrate of *Acrophialora nainiana* and *Stachybotrys atra* are effective (Dasgupta and Mandal, 1989).

2.6 Weed Pests

Weed is an unwanted plant in a crop field. They compete with the main crop for light, air, moisture, nutrients and thus reduce crop yields up to 37 per cent (Varshney, 2009). Distribution of weed flora varies with soil, agro climatic zones, season and particular crop species.

Since the cucurbits are vine crops and mounted upon pandals or trellis they can smoother the weed growth by their spreading vines or developed crop canopy particularly at later stages of plant growth. But in some cases poor crop stand or defoliation of late season crop growth by pest attack can pose scrious threat at reproductive stage. However, management of weeds of cucurbits at later part of crop growth is also important for better air circulation for the crop and easy harvesting. Winter squash and pumpkin produce bigger foliage cover than others. During early vegetative growth i.e at 4-5 leaf stage the pits or ridges should be kept weed free. The minimum weed-free period in cucumber, squash, and other cucurbit crops has been estimated as the first 4 to 6 weeks after planting (Noble, 2009).

In India mechanical or chemical method of weed control are generally followed for cucurbits. But integration of several management practices is viable option for suppression or prevention of weed growth in cucurbit vegetables.

Common weed flora of cucurbits

Cucurbits are basically warm season crop. But due to development of thermo insensitive varieties this type of crop is grown throughout the year which has in turn invited some more weeds in the fields of pumpkin, gourds, etc. The commonly found weeds in the plots of cucurbits are given hereunder.

2.6.1 Broad leaved weeds

2.6.1.1 Pig weed: *Amaranthus viridis* Hook. Family: Amaranthaceae

It is a broad leaf weed and commonly found during kharif season in the plots of pumpkin, pointed gourd, bitter gourd, ridge gourd etc. This annual herb can grow up to a height of 100 cm. with glabrous to pubescent stem and leaf and *catkin* like *cymes* of densely packed flowers.

2.6.1.2 Spiny amaranth: *Amaranthus spinosus* L. Family:Amaranthaceae

It is another *kharif* season obnoxious weed of cucurbits in the tropics and subtropics, growing up to a height of 5 feet. Leaves are alternate and ovate. One plant can produce 235,000 seeds. Small, green colored male and female flowers are borne on the same plant. Fruit is ovoid and contains compressed, shiny, tiny, dark red to black seeds.

2.6.1.3 Black night shade: *Solanum nigrum* Linn. Family: Solanacae

Black night shade grows as weed in the dry parts of India and other parts of world (Akilan *et al.*, 2014). This is a *kharif* season broad leaf weed found in the plots of cucurbits. The plant grows up to one meter height with erect, glabrous or sparsely pubescent stem and leaves are ovate, glabrous and thin, margins toothed, tapering into the petiole. The plant produces small drooping flowers and yellow or black colored berry like fruits and minutely pitted seeds. Flowers are small, white, borne in drooping, umbellate 3-8 flowered cymes. Black night shade acts as host for cucurbit powdery mildew fungus (*Erysiphe cichoracearum*).

2.6.1.4 Common purselane: *Portulaca oleracea* Linn. Family: Portualacaceae

It is an annual glabrous herb with prostrate and succulent stem. Leaves are spatulate, flattened, apex round nearly truncate. Flowers are 3-10 mm in diameter and yellow in colour. Fruits are capsules ovoid, 4-9 mm diameter. Seeds black or dark brown, orbiculate or elongate, flattened, 0.6-1.1 mm in size; surface cells are smooth, granular, or stellate with rounded tubercles. This weed also act as host of cucumber mosaic virus (CMV).

2.6.1.5 Chick weed: *Stellaria media* (Linn.) Vill. Family: Caryophyllaceae

This is winter annual broad leaf weed which grows up to 12 inches height. Stems are prostrate, leaves are opposite, produces small star shaped flowers. Fruit is a capsule which contains many tiny reddish brown seeds (Whitson *et al.*, 2000). Chick weed acts as host plant for melon aphids which attacks most of the cucurbits (Hazra and Som, 2015).

2.6.1.6 Wild morning glory: *Convolvulus arvensis* Linn. Family: Convolvulaceae

This is a winter perennial plant which produces smooth, slender, glabrous stems of 2-4 feet long. Plant produces white or pink color funnel or trumpet shaped flowers. Fruit are light brown, rounded and contains 2 seeds which remain viable in soil for long period. The plant produces deep, perenniating tap roots with several lateral roots that spread radially and act as a method of vegetative reproduction (Tenaglia, 2006). Wild morning glory acts as host plant for aphid in water melon, pumpkin, ridge gourd,etc. (Hazra and Som, 2015).

2.6.2 Sedges

2.6.2.1 Nut grass: *Cyperus rotundus* Linn. Family: Cypraceae

This is a perennial weed found in many field of vegetable crops like cucurbits due to its unique biological and physiological properties (Islam *et al.*, 2009). The plant produces narrow linear leaves. Inflorescence is a simple or compound umbel, bearing short spikes of 3-10 spreading, red-brown spikelets. It reproduces mostly through tubers (Sharma and Gupta 2007, Lati *et al.*, 2011). Keeley (1987) reported that yield losses could reach 43% in cucumber (*Cucumis sativus* Linn.) due infestation of nut grass.

2.6.2.2 Lambs quarter: *Chenopodium album* Linn. Family: Chenopodiaceae

It is a *rabi* season annual, 0.3- 3 mt tall with green, red or purple stems. Leaves are alternate, simple, green above and mealy-white below. Flowers are minute and are arranged in densely clustered panicles. Fruits utricle, seeds round, compressed, black and shining.

2.6.3 Grasses

2.6.3.1 Bermuda grass: *Cynodon dactylon* (Linn.) Pers. Family: Poaceae

It is one of the commonly found obnoxious perennial weed in cucurbits. It produces fibrous root system with above ground stolons and underground rhizomes, with narrowly linear or lanceolate leaves arising from upright stem branches.

2.6.3.2 Barnyard grass: *Echinochola crusgali* Linn. Family: Poaceae

It is a kharif season annual, 30 cm to 2 m tall, culms cylindrical, glabrous, leaves are alternate, flat, glabrous, 30-50 cm long. Inflorescence is a panicle green or purple, 5 - 25 cm long. Fruit is a caryopsis.

2.6.4 Management of weeds

Due to higher cost and crisis of human labour no single method of weed management is viable and cost effective. Judicious management of weeds needs long term strategy and integration of different sustainable methods because effective weed management is critical to maintaining agricultural productivity (Ahmed *et al.*, 2010, Verma, 2014).

Among different weed flora in cucurbits control of rhizomatous perennials like nut sedges, Bermuda grass, etc. is a difficult task. Vine weeds like morning glories, hedge bindweed (*Calystegia sepium*) and field bindweed (*Convovulus arvensis*) twine around and make the harvesting difficult. Many non-chemical weed management methods are common in farming practices. These practices are of increasing importance due to consumers' concerns about pesticide residues, potential environmental contamination from pesticides, and unavailability of many older herbicides (Masiunas, 2000). Management of different weed flora in cucurbits need selection and integration of different control methods which may be summarized as below.

2.6.4.1 Preventive methods

a) **Use of clean seed**

As many weed seeds of many harmful weeds are mixed with the main crop and these seeds should be separated or cleaned before sowing to avoid the contamination. For this seeds should be purchased from some authentic sources. In cucurbits seeds of wild cucurbits (*Cucurbita spp.*) get easily mixed with that of cultivated cucurbits and regarded as objectionable (Jaya Kumar and Jagannathan, 2003).

Weeds of cucurbit family, such as bur cucumber (*Sicyos angulatus*) and wild gourd (*Cucurbita pepo*) acts as alternate host for squash bug (*A nastatristis*) (Adam, 2006) and should be removed. Plants of wild gourd should be uprooted or eradicated anywhere within a quarter mile of seed production of *C. pepo* squash or pumpkin varieties to avoid cross pollination with the wild relative that can introduce genes for intensely bitter and toxic cucurbitacins (Connolly, 2005).

b) **Clean farm implements**

Seeds or propagules of many weeds stick to the implements used in the previous crops. So proper cleaning or disinfection of these implements are needed to avoid the spread of the particular weed.

c) **Well decomposed manure**

For cultivation of cucurbitaceous vegetables thoroughly decomposed FYM should be used as many weed seeds remain viable in raw or undecomposed manure. Some weed seeds like *Abutilon theophrasti* Medic, *Convolvulus arvensis and Melilotus* sp showed some viability, 2%, 4% and 22%, respectively even after storage in cow manure for one month.

d) **Clean irrigation channels**

Irrigation water carries soil and weed seeds to a crop field. So proper cleaning of irrigation channels, use of micro irrigation or weed seed screen filter is needed for prevent the dispersal of weed seeds in the crop field.

e) **Implementation of Weed laws**

Proper enactment or imposition of weed law or quarantine is needed to check the entry or spread of noxious and pernicious weeds such as *Striga* sp, *Orobanche* sp, *Parthenium hysterophorus, Eichhornia crassipes,* , *Lantana camara* etc. It could be both inter-state and inter-country movement.Weed law exists only in Karnataka in India, which declares *Parthenium hysterophorus* as a noxious weed. It should be enacted across states and countries and should include most of the noxious weeds.

2.6.4.2 Curative methods

a) **Mechanical method :** Removal of weeds physically or with mechanical power is used to kill or pull out weeds from the crop field. Most commonly used mechanical weed control methods, such as hoeing, tillage, harrowing, torsion weeding, finger weeding and brush weeding, are used at very early weed growth stages (Singh, 2014, Kewat, 2014). In cucurbits weed pose problem specially during early vegetative growth i.e. spreading of vines. Hence, mechanical method of weed control is very effective for those crops.

i) **Hand pulling/hand weeding** : Hand weeding is done by small implements like khurpi, sickle, etc. to remove them along with roots out of crop field. This loosens soil surrounding the crop rhizosphere and enhances crop growth and yield. In cucurbits hand weeding is done only during initial stages of growth and development as fast growth of vines of all the cucurbits can smoother weed population at later phase. However, hand weeding is time and labour consuming.

ii) **Thermal method :** Here weeds are controlled through burning, flaming, steaming or freezing (Ascard *et al.*, 2014). Burning is practised mainly in fallow land for towards non-selective control of weeds or wild vegetation along with destruction eggs, larvae of insect pests or soil borne pathogens whereas flaming is applied both selectively and non-selectively.

iii) **Mulching :** Mulching is one of the possible ways to control weeds without using herbicides (Verma and Singh, 2008, Awasthy *et al.* 2014). Polyethylene mulch has been shown many times to increase cucurbit yield and earliness. Mulches may be organic (straw, bark, wood ash, compost, etc.) or synthetic (plastic). Black plastic mulches are commonly used and effective for better early season growth of warm-season crops such as tomatoes, muskmelons, watermelons, and peppers and thereby making the crop more tolerant or suppressive against weeds (Bhullar *et al.*, 2015). However, viny weeds in cucurbits can grow out of the planting holes of plastic mulched crops and the nut sedges often puncture and grow through the plastic itself.

iv) **Soil solarization :** Soil solarization is an effective method of weed management where moist soil is covered by polyethylene film (usually black or clear plastic sheet) to trap solar radiation and cause an increase in soil temperature for several weeks to levels that kill weeds, weed seeds, plant pathogens, and insects for economic crop production (Singh, 2014). White polyethylene cover is better than black one.

b) **Cultural method :** Cultural methods provide competitive advantage to crop against weeds by reducing weed establishment (Singh, 2014) and through selective stimulation, facilitating faster crop growth to smother weeds (Das *et al.*, 2012).

i) **Selection of crop species and variety :** A crop with vigorous growth habit, fast growth and large canopy cover can smoother the weed growth rapidly. Most of the cucurbits with their vast canopy cover can suppress the weed growth very efficiently. Similarly choice of suitable variety or cultivar can prevent or minimize the weed flora very effectively. The cultivar with faster seedling emergence, canopy establishment, early fast growth, maximum number of leaf, tall stature and more tillering capacity have better competitive ability against weeds (Ahmed *et al.*, 2010, Bhan *et al.*, 2012).

ii) **Crop rotation :** It is another important weed management practice for cucurbits. It is highly effective against parasitic weeds such as *Orobanche spp.* and other obnoxious weeds like *Echinochloa colona ,Phalaris spp.* Growing cool season vegetables and weed-competitive summer cover crops (e.g., snap pea–buckwheat–fall broccoli) during the year prior to a cucurbit crop help to disrupt summer annual weed life cycles.

iii) **Sowing or planting time :** Sowing or planting of crop at right time shows more competitiveness over weed flora than late sowing or planting. Early planting provides a competitive edge to adapted crop cultivars (Sindhu *et al.*, 2010) because crop emerged before the weeds and therefore the weeds did not receive sufficient sun light for their emergence and growth.

iv) **Stale seedbed technique :** This technique is followed for many vegetable crops including cucurbits. Here the seed bed after preparation is irrigated and left unsown to come up the weeds which later on are killed by light tillage operation or by using non residual herbicides just before or after planting, but before crop emergence. The crop is planted with minimum soil disturbance to avoid exposing new weed seed to favourable germination conditions.

v) **Intercropping :** Intercropping or growing more number of crops together with different duration or growth habit is another cultural method of weed control. Intercropping, preferably spreading types of crops, like cucurbits produces a faster and denser ground cover suppresses weed growth and reduces erosion (Giri *et al.*, 2006).

vi) **Zero tillage :** Zero tillage or no tillage has shown positive results in controlling several weeds in cucurbits. Walters *et al.* (2007) indicated that a winter rye cover crop alone would provide some but not sufficient, season-long redrootpigweed and smooth crabgrass [*Digitaria ischaemum* (Schreb. *ex* Schweig.) Schreb. *ex* Muhl.] control for cucumber grown in no tillage condition. The use of herbicides along with cover crops sometimes play vital role in the management of weeds in no tilled pumpkin production. Several studies have all indicated that no tillage and conventional tillage produce comparable pumpkin yields when sufficient weed control is achieved in no tillage production systems (Walters *et al.*, 2008).

vii) **Allelopathy :** Allelopathy is any direct or indirect effect that one plant exert upon another when they are grown in combination. The exploitation of crop allelopathy against weeds may be useful to reduce issues related to the use of herbicides because in recent years, allelopathic suppression of weeds is receiving greater attention (Inderjit *et al.*, 2005). Among cucurbits cucumber showed allelopathic effect to *Echinochloa crusgalli.*

c) **Biological method :** Biological control of weeds includes different living agents like predators, parasites, fishes, snails or botanicals to control the weeds. Pioneering works on biological control of weeds was carried in India for control of *Parthenium hysterophorus* (Kumar and Ray, 2011). Information on biological control of weeds for cucurbits or other vegetable crops is lacking in India or abroad.

d) **Chemical method :** Control of weeds through synthetic chemicals or herbicides offers a good strategy in cucurbits. But the heavy reliance on chemical weed control is nowadays considered objectionable (Das, 2008) as extensive use of herbicide causes hazard on food safety and environment. Continuous use of synthetic herbicides poses resistance among different weed flora. Mode of action, time of application, etc. varies with different herbicides vis vis different cucurbits. Chemical control of weeds of different cucurbits is given in Table 1.

Table 13. Chemical control of weeds in different cucurbits

Crop	Control measure	Time of application
Musk melon	Alachlor @ 2.0 kg a.i. /ha or fluchloralin @ 1.2 kg a.i./ha	Pre sowing
Watermelon	Butachlor @ 2.0 kg a.i. /ha	Pre sowing
Bottle gourd	Alachlor @2.5kg a.i. /ha or fluchloralin @ 2.0 kg a.i./ha or Butachlor @ 2.5 kg a.i. /ha	Pre sowing
Bitter gourd	Alachlor @2.5kg a.i. /ha or fluchloralin @ 0.75 kg a.i./ha or Butachlor @ 2.5 kg a.i. /ha	Pre sowing
Pumpkin and summer squash	Butachlor @ 2.0 kg a.i. /ha	Pre sowing
Ridge gourd	Alachlor or Butachlor @ 2.0 kga.i./ha	Pre sowing

2.7 Host Plant Resistance in Cucurbits Against Biotic Stresses

Resistance is any inherited characteristic of a host plant that lessens the effects of insect infestation. From the evolutionary point of view, such characters are a pre-adaptive trait that have allowed the organism to overcome the pressures of insect pests and other biotic stresses and thereby increases the organism's chances of survival and reproduction (Pedigo, 2002).

The technology of plant breeding for pest resistance is the secondary objective of a total breeding programme in addition to yield and quality. It focuses on any types of pests but usually emphasis is given on pathogenic microorganisms, nematodes and arthropods. Host plant resistance to insect pests and diseases is an underutilized pest management strategy in vegetable production. Increased pressures to reduce pesticides and changes in technology now increase the economic viability and probable role of host plant resistance in vegetable pest management. This is reflected in the relatively recent release of several insect resistant varieties and breeding lines (Sanford *et al.*, 1994). Like other vegetables there is ample opportunity to exploit this technology in managing several biotic stresses in cucurbits.

Resistance sources are generally present in land races and wild relatives. Resistance to downy mildew (*Pseudoperonospora cubensis*) is reported in snap melon (*C. melo* var. *momordica*), resistance to fruit fly is reported in *C. callosus* etc. Most of the resistant varieties in cucurbits have been developed by simple selection.

Traditional development of resistant cultivars

Development of plant resistance involves active participation of specialists from different concerned disciplines. The programme requires input from plant geneticists, plant breeders and entomologists. The actual crossing, genetic analysis and evaluation of agronomic characteristics are the responsibilities of plant scientists and entomologists contribute by identifying the bases of resistance. The scientists have to identify the sources of resistance in a resistance cultivar and transfer those traits to another cultivar through crossing. After initial crossing the subsequent progeny of the cross are evaluated for the resistant trait. This process is sometimes repeated several times to become confirm regarding resistance. The plant breeders also can identify the resistance genes and measure their degree of expression in the newly developed cultivars.

Complementary functions of host plant resistance

Use of resistant plant is one of the most important key components for successful IPM system. It has even greater potentiality than any other single tactic or strategy for pest suppression. In theory as well as in practice, the compatible and complementary role of host plant resistance is in concert with the objectives of IPM. Biological control is not compatible with insecticidal control of crop pest. Again, host plant resistance is not subject to the natural calamities as are chemical and biological control. In addition to these:

a) Host plant resistance is not density dependant as biological control.

b) In endemic situation and typical climatic condition host plant resistance is the only effective means of pest management.

Successful uses of insect resistant cultivars

Although major development with regard to plant resistance to insects have taken place in grain and forage crops including wheat, rice, corn, sorghum, sugarcane, beans, barleys and alfalfa, but considerable progress with varying degree of success has also occurred in vegetables. As per a survey of G. F. Sprague and R. G. Dahms in United States, more than 100 cultivars resistant to more than 25 different insect species had been released to commercial production by 1972 (Pedigo, 2002). In India also a good number of insect resistant cultivars of different vegetables have been developed and commercialized.

Use of host plant resistance in pest management

In insect pest management the use of host plant resistance is obviously a prophylactic measure. Most of the resistant cultivars developed so far are for severe pests where significant insect pest populations are probable. Use of resistant cultivar reduces the carrying capacity of that particular agro-ecosystem and ultimately drops down general equilibrium position of the pest population. However, selective pressure is placed on the pest population where resistant cultivars are grown. But in case of tolerance cultivars the carrying capacity of the agro-ecosystem does not get reduced and thereby no selective pressure in placed on the pest population.

Host resistance as the sole pest management tactic

Host plant resistance have been used as the only management tactic where irrepairable losses are usually taken place and such losses could not be compensated by applying curative measure like the use of pesticides. Other preventive measures usually lack effectiveness. Considerable successes in using host plant resistance as the sole management technique against several insects have been achieved. But the threat of resistance breaking biotypes, continual monitoring of the pest's status is of utmost necessity. Dynamic breeding programme and development of transgenic plant variety to be given special stress for timely supply of new resistant varieties when they are in need.

Host plant resistance integrated with other techniques

Long tasting solution to insect pest management is depends on successful intellectual formulation and implementation of several harmonious tactics. Host plant resistance can have the ability to serve as an excellent tactic as it is usually compatible with almost all other pest management techniques including biological control.

Integrating host plant resistance and biological control

Selection pressure imposed by the natural enemies should result in the magnification of plant resistance (Van Emden, 1990). Host plant resistance and biological control are very much compatible with each other. Application of these two pest management strategies brings together unrelated mortality effects on the pest population. Thus, they reduce the pest population's genetic response to selection pressure either from host plant resistance of natural biological control. The rate of insect adaptation on a resistant variety is lower at a given level of insect population suppression when that suppression is achieved by the combined action of plant resistance and natural enemies than by a high level of plant resistance that leads to the development of biotype (Panda, 2003). Host plant attributes such as spatial and temporal distribution, architecture and resistance against herbivorous insects can affect organisms like natural enemies at the third trophic level. However, until recently a very few studies have examined the effects of plant morphology, physiology and plant allelochemicals on the biology of predators and parasites.

Some studies revealed that morphological characteristics and plant defense chemicals have adverse effects on the predators. Increased trichome density on leaves reduced the ability of the parasites *Trichogramma* and *Chrysoperla* to find out and parasitise their hosts. So, these tritrophic levels of interactive systems should not be ignored rather should be fully realized while formulating resistant breeding programme.

Integrating host plant resistance with microbials

Bt endotoxins applied as insecticides are apparently generally compatible with HPR. Efficacy of the endotoxins is equal or greater on resistant varieties as compared with susceptible varieties (Hare, 1992). There is evidence that allelochemicals associated with resistance can potentiate *Bt* endotoxins (Ludlum *et al.*, 1991, Trumble *et al.*, 1991, Meade and Hare, 1993). Almost all the examples are on vegetables or concern allelochemicals prominent in vegetable crops. There is a potential for reduced efficacy of the *Bt* toxin if less is ingested during insect feeding on a less preferred crop, but this has not yet been demonstrated (Meade and Hare, 1993).

On the other hand, plant allelochemicals or plant resistance can have a negative effect on the toxicity of insect pathogens including pathogenic fungi and nuclear polyhedrosis viruses (NPV) (Hare and Andreadis, 1983, Felton *et al.*, 1987, Felton and Duffey, 1990). Possibly that resistant plant and insect pathogens may be antagonistic and hence should be examined when these elements are combined in IPM.

Integrating host plant resistance and chemical control

Use of resistant cultivars invites less insect-pest population on the crop and hence integration of host plant resistance with chemical pesticidal control usually results in very low input of pesticides as compared to their counterpart of susceptible cultivars. Frequency of pesticide application gets reduced. Sometimes sub-lethal dosage of pesticide may act as repellant of insect-pest and thus protect the crop. In a two locus model, Gould et al. (1991) illustrated that insect resistance against pesticides can also be managed in the field by using insecticides having insect repellant properties coupled with the repellant properties of the host plant. The model predicts that insecticide formulations containing non-insecticidal compounds with repellant properties (neem and other botanicals) could lower the rate of insect adaptation to an insecticide. Insects lack the apparatus to respond to most insecticides, but they have evolved the ability to respond to volatile secondary compounds. Therefore, the selective pressure due to insecticide treated populations of insects on plants without *antixenosis* is likely to be much stronger than that occurring in treated populations faced with a resistant host variant with *antixenosis* resistance (Pluthero and Singh, 1984).

Insect resistance traits of a resistant cultivar may variously affect the efficacy of pesticides by mechanically affecting coverage, or through physiological effects on the target pest. These include induction of detoxifying enzymes in insect guts (Ahmad *et al.*, 1986), changes in feeding rates affecting pesticide ingestion (Abro and Wright, 1989), reduced body size or general vigor increasing insecticide susceptibility.

Integrating host plant resistance and cultural control

Cultural Control is obviously a powerful weapon in IPM to suppress the insect pests in an Agro-ecosystem. It has not that ability to keep the pest population below economic threshold level but may aid in reducing losses due to pests. Host plant resistance, holistic plant health care system in concert with cultural control can effectively reduce the need for pesticides vis-à-vis lowering the cost of cultivation.

Options for cultural practices are:

a) Physical barrier or trap cropping.

b) Adjustment of planting time.

c) Optimum use of balanced fertilizer and water.

d) Proper cropping pattern.

Introduction of host plant resistance in vegetable cultivation

In the days of modernization of agriculture there are a number of open pollinated and F_1 hybrids resistant to insect pests in vegetable crops developed by conventional plant breeding methods. As discussed earlier, the genetic resistance to insect pests may be due to non-preference, antibiosis, tolerance or a combination of these factors. Some important cultivars/varieties tolerant to insect pests in various vegetables developed in India are mentioned in Table 14.

Table 14. Resistant cultivars of vegetables against pest and diseases.

Crops	Insect pests/ diseases	Resistant varieties	Author(s)
Pumpkin	Fruit fly	Arka Suryamukhi	Satpathy *et al.*, 1998
Round gourd	Fruit fly	Arka Tinda	Satpathy *et al.*, 1998
Cucumber	Powdery mildew	Phule Shubangi	Satyagopal *et al.*, 2014
	Anthracnose, Powdery mildew, Downy mildew, Angular leaf spot	Poinsettee	Choudhary and Fageria, 2002
Bottle gourd	Blossom end rot	Arka Bahar	Satyagopal *et al.*, 2014
	Anthracnose, Powdery mildew, Downy mildew	N. Shishir (NDBG-202)	Satyagopal *et al.*,2014
Bitter gourd	Downy mildew	Phule Green Gold	Satyagopal *et al.*, 2014
Melon	Powdery mildew	Arka Rajhans	Rai *et al.*, 2008a
	Downy mildew	Punjab Rasila	Rai *et al.*, 2008a
	Cucumber Green Mottle Mosaic Virus	DMDR-2DVRM-1DVRM-2	Rai *et al.*, 2008a
Watermelon	Anthracnose, Powdery mildew, Downy mildew (Multiple resistance)	Arka Manik	Rai *et al.*, 2008a
Muskmelon	—	Home garden, Perlita, Planers Jumbo	Choudhary and Fageria, 2002
Summer squash	Powdery mildew and Cucumber mosaic virus	Kaddu-1, Punjab Chappan	Choudhary and Fageria, 2002

3

Abiotic Stresses

Tremendous genetic diversity occur within different crops of this family with respect to their adaptability. Some cucurbits are grown in tropical, subtropical or arid deserts whereas other prefer to grow in temperate conditions. Harvested fruits and vegetables can be potentially exposed to numerous abiotic stresses during production, handling, storage and distribution (Hodges, 2003). However, when the abiotic stress is moderate or severe, quality losses almost always are incurred at market (Toivonen, 2003). Annually about 42% of the crop productivity is lost owing to various abiotic stress factors (Oerke *et al.*, 1994). One of the key factors for abiotic stress in crop plants is the ill effect of global warming which may cause reduced precipitation, less snow pack, and earlier snow melt, leading to drought conditions. In addition about 20% of the world's irrigated lands are affected by salinity (Zhu, 2001), a situation worsened by climate changes.

Global warming increases the rates of evapo-transpiration and crop requirement of irrigation water, which may bring more salt into the soil in arid and semiarid regions. High salinity limits crop production in 30% of the irrigated land in the United States (Kumar *et al.*, 2004). Like other vegetables this group of vegetables are also sensitive to different abiotic stresses.

3.1 Nutritional Disorders

3.1.1 Blossom-end rot

This disorder of watermelons has been studied in Iraq and in Italy (Cirulli and Ciccarese 1981). Cylindrical fruited cultivars are especially susceptible to this disorder.

Causal factor

It is primarily a nutritional disorder associated with insufficient calcium uptake and alternating periods of wet and dry soil. This lack is the result of slowed growth and damaged roots caused by any of several factors.

- Extreme fluctuations in soil moisture.
- Rapid early-season plant growth followed by extended dry weather.
- Excessive rain that smothers root hairs.
- Excessive soil salts.
- Cultivating too close to the plant.

Symptoms

The main symptom of this disorder is the sunken area at distal end of the fruit. By harvest time about 75% of the crop may be affected. Although, this disfigurement does not progress after harvest. It constitutes a serious market blemish.

The first fruits of the season are the most severely affected. As the name implies, the disorder always starts at the blossom end, though the rot may enlarge to affect half of the fruit. Moldy growths on the rotted area are caused by fungi or bacteria that invade the damaged tissue.

Management

a) Growing spherical fruited cultivars which are not generally susceptible to this disorder (Cirulli and Ciccarese, 1981).

b) Maintain uniform soil moisture by mulching and proper watering.

c) Avoid high-ammonia fertilizers and large quantities of fresh manure.

d) Lime deficient soils may require amendment with gypsum (calcium sulphate).

e) Frequency of irrigation may need to be increased to permit efficient uptake of minerals.

3.1.2 Pillow

It is a fruit disorder of processing cucumber.

Causal factor

It takes place due to low calcium level in the tissue.

Symptoms

In this disorder, an abnormal white styofoam like porous textured tissue is formed in the mesocarp of the fleshy harvested fruits. Vascular tissue with some pillow areas may collapse and become necrotic.

3.1.3 Interveinal chlorosis

Causal factor

It occurs due to calcium and magnesium deficiency.

Symptoms

Stunted plant growth with interveinal chlorosis.

Management

a) Use balanced fertilizer.

b) Calcium and magnesium fertilizer may be applied.

c) Apply chelated form of iron.

3.1.4 Molybdenum deficiency

The disorder is commonly observed in cucumber.

Causal factor

Deficiency of the micronutrient molybdenum.

Symptoms

Stunted growth with whitish interveinal chlorosis accompanied by marginal leaf burn.

Management

a) Soil or foliar application of sodium or amonium molybdate.

3.1.5 Yellowing and scorching of leaves

Causal factor

Deficiency of potassium causes yellowing and scorching of older leaves of the plant.

Symptoms

Symptoms begin to appear at the margins of the leaf and spread in between the veins towards centre. A brown scorch develops in the yellow areas and spreads until the leaf is dry and papery. Large areas of tissue around major veins remain green until the disorder is well advanced.

Management

a) Foliar spray of KCl 1% at weekly interval.

b) Potassium fertilizers are best in managing the disorder if incorporated in the soil before planting.

c) Fertigation or drip feeding can also be used to treat a deficient crop.

3.1.6 Leaf yellowing

Causal factor

Magnesium deficiency causes yellowing of older leaves.

Symptoms

The symptom begins to appear in between the major veins, which retain a narrow green border. A light tan burn will develop in the yellow regions if the deficiency is severe. Fruit yields are reduced.

Management

a) Incorporation magnetite (300 kg/acre) or dolomite (800 kg/acre) into deficient soils before planting may be recommended.

b) Fortnightly foliar sprays of $MgSo_4$ (2 kg/100 L) at high volume (500-1000 L/acre) is also helpful in managing the problem.

3.1.7 Boron deficiency

Symptoms

Distortion of newer leaves takes place (in severe cases the growing points also die) and a broad yellow border appeared at the margins of the oldest leaves. Young fruit can die or abort. Fruit abortion rate become high. Stunted development and mottled yellow longitudinal streaks, which develop into corky marking (scurfing) along the skin.

Management

a) Foliar spray of 0.2% Borax at forthrightly interval.

b) Application of 10 kg Borax per hectare in case of deficient soil before sowing will prevent boron deficiency.

3.1.8 Iron deficiency

Symptoms

Iron deficiency causes a uniform pale green chlorosis of the tender leaves. All other leaves remain dark green. Initially the veins remain green, which gives a net-like pattern. In severe cases, the minor veins also become fade. The leaves may eventually burn, especially if exposed to strong sunlight.

Management

a) Good drainage and soil aeration favour iron availability.

b) Foliar spraying of iron sulphate (150 g/100 lt) can be used to treat the affected plants.

3.1.9 Manganese deficiency

Symptoms

The veins of middle to upper leaves of manganese-deficient plants appear green against the mottled pale green to yellow of the blade.

Management

a) Spray the foliage with $MnSO_4$ @ 0.1% (100 g/100 lt water).

3.1.10 Salt injury

Causal factor

This problem is associated with excessive soluble salts specially in green house cucurbits.

Symptoms

Plants become dark green in the early stages, but rapidly develop marginal yellowing and necrosis of older leaves.

Management

a) Careful fertilization and watering reduce the problem.

b) Watering container-grown plants to runoff will prevent salt build up.

3.2 Environmental Disorder

3.2.1 Chilling injury

This is commonly seen in melons and cucumber.

Causal factor

Chilling injury mainly seen when fruits are stored at temperature below 12°C. The critical temperature below which chilling injury can occur is between 10°C to 14°C for water melons, melons which are less susceptible and their cultivars differ substantially.

Symptoms

This disorder may take the form of reddish brown patches on the skin of fruits (Picha, 1986, Lipton and Wang, 1987). Ripening may also be prevented and taste and texture impaired. Chilling injured tissues shows an increased tendency to get decayed. Multiple lesions are developed on the surface of fruits. Typical post chilling infections are *Alternaria* rot and *Cladosporiunm* rot.

Management

a) Sensitivity to chilling injury can be reduced by plastic film wraps, pre-shipment treatment with ethylene or pre-shipment conditioning for a few days at moderate temperature.

b) In field condition, mulching may be practised.

3.2.2 Solar injury

Causal factor

This injury may result from UV ray or heat or a combination of both (Lipton, 1977). Solar injury is common in melons. This problem is serious in crops with insufficient foliage cover growing in hot and arid climate.

Symptoms

Patchy ground colour or 'bronzing' and discolouration appear on fruit. In winter melons the affected area may be brown or blotched. Severely injured tissues become sunken or wrinkled. Flesh quality may be impaired. In hot arid growing areas solar injury can be a problem in crops where there is inadequate foliage cover to shade the maturing fruits. Harmful exposure to the sun may also occur at harvest time if fruits are not brought under cover promptly. In this situation flesh temperature can rise several degrees above the ambient temperature that sometimes may result into irreversible damage.

Management

a) Apply white wash to the melons so that sun light and heat are reflected (Lipton 1977).

b) Providing shade after harvest and transport harvested melons without delay.

3.2.3 Flood damage

Almost all the cucurbits are susceptible to prolong waterlogged condition.

Causal factor

Anaerobic condition due to continuous flooding.

Symptoms

The symptom often appears as nutrient deficiencies or a generalized yellowing. Prolonged exposure to flooded soils lead death of plants.

Management

a) Planting in raised beds will improve drainage and will reduce the problem.

3.2.4 Measles

This is common in smooth-skinned melons and cucumbers.

Causal factor

This is associated with environmental conditions favoring guttation. The guttation droplets develop high concentrations of salts which burn the epidermis. Measles spots occur where a guttation droplet had formed.

Symptoms

Small brown spots are scattered over the surface of the fruit. The spots do not penetrate beyond the outer epidermal layers of the fruit. These spots also may occur on leaves and stems.

Management

a) Reduce irrigation frequency and duration as fruit approach maturity.

3.2.5 Vein-tract browning

It is a post harvest disorder of melons.

Causal factor

Ageing, accelerated by exposure to the sun during maturation and to high temperature after harvest (Lipton, 1977).

Symptoms

Darkening of longitudinal grooves of the fruits.

Management

a) Storage in high humidity can reduce this problem.

3.2.6 Rind necrosis

Generally occurs in either cantaloupe or watermelon.

Causal factor

Probably due to environmental stress like drought.

Symptoms

Dead, hard, dry reddish-brown to brown spots or patches of tissue appear on the fruit rind. In watermelon, symptoms are not visible from the outside and are rarely found in the flesh. In cantaloupe, dead tissue may extend into the flesh of the fruit.

Management

a) Avoid drought stress in melon.

3.2.7 Bitterness

This is a common problem for most of the cucurbits due to production of chemicals called cucurbitacins.

Causal factor

Higher levels of cucurbitacin triggered by environmental stress, like high temperatures, wide temperature swings or too little water. Uneven watering practices, low soil fertility and low soil pH, over mature or improperly stored cucurbits may also develop a mild bitterness.

Symptoms

Bitterness of fruits.

Management

a) Proper selection of varieties with optimum supply of irrigation and fertilizer.

3.2.8 Drought stress

Cucurbits are particularly sensitive to drought. Fruits are typically 85% to 90% water and can suffer under drought condition. Pumpkins often produce long vines with many leaves and can transpire large quantities of water during hot summer days. Severe drought stress affects fruit development, resulting in unmarketable produce or tapered at the blossom end. Pumpkins become soft and wrinkled under drought. In addition, drought-stressed pumpkins fail to gain appropriate size, which affects yields. A loss of foliage during drought will also result in sunburn of the fruit.

Management

a) Irrigation to be given as and when necessary.

3.3 Others

3.3.1 Premature senescence

This type of problem is found in watermelon.

Causal factor

This disorder occurs in presence of ethylene.

Symptoms

Watermelons are usually harvested when already ripe or very nearly so and do not have much capacity for improvement in flesh quality. On the contrary, once the fruits are harvested it is not long before there is decline in colour, texture and flavor of the ripe flesh. Such changes like fading, softening and development of off flavours are accelerated by exposure to ethylene (Risse and Hatton, 1982).

Management

a) Segregation of watermelons during shipment and storage from banana, tomato, apples etc. because these fruits produce substantial quantities of ethylene.

3.3.2 Mis-shapen fruits

The fruits may be crooked or constricted and it is a common problem of cucurbits.

Causal factor

It is a pollination problem. Field grown cucurbits usually require pollination for fruit development. Due to lack of pollinating agents this disorder may happen.

Management

a) Release of one bee per plant or 2.5 strong and active hives per hectare. In case of high density planting five hives per hectare need to be introduced.

b) Insecticides should not be applied near the hives or in the field when the bees visit flowers especially from sunshine to early afternoon.

3.3.3 Fertilizer burn

This injury is found in all cucurbits. Although all vegetables can be affected by fertilizer burn, cucurbits are particularly very much sensitive because they do not have a thick waxy cuticle on their leaves. Therefore, they do not shed water as in some other vegetables, such as onions or the *Brassica* species.

Causal factor

This situation occurs when chemical fertilizers (composed of salts) are applied at high concentrations.

Symptoms

Symptoms include a generalized burned appearance or flecking on plant.

Management

a) Avoid foliar feeding if possible.

b) Growers should plan on providing all the necessary fertilizers for their crops through fertigation or soil applications.

3.3.4 Hollow heart

It is the formation of a hollow cavity inside some cucurbit fruit mostly watermelon. Uneven visible hollow heart makes fruit unmarketable.

Causal factor

This disorder can result from a number of factors like genetic poor pollination followed by rapid fruit growing conditions (too much fertility, water and high temperatures).

Management

a) Avoid varieties with a tendency to exhibit hollow heart.

b) Ensure that boron levels in the soil are adequate; however, be careful not to overfertilize.

c) Follow recommended plant spacing, fertilizer dose and avoid erratic irrigation.

3.3.5 Leaf silvering

It is a physiological disorder of summer squash (*Cucurbita pepo*).

Causal factor

This disorder occurs due to moisture scarcity.

Symptoms

The leaves become silver coloured and contain less chlorophyll; photosynthesis is hampered in the silvered leaves.

Management

a) Maintenance of optimum moisture in field.

3.3.6 Unfruitfulness

This is a common problem in pointed gourd which is a dioecious cucurbit.

Causal factor

It occurs due to poor or no pollination.

Symptom

Poor or no fruit set observed. A common problem is met with where pistillate flowers in female plants are shed due to lack of pollination and fertilization. In some cases, ovary of the unfertilized flower may flow a bit due to parathenocarpic stimulation which also abscise after a few days.

Management

a) Maintain male: female in 10-12:100 ratio to ensure adequate pollination and fruit set.

b) Hand pollination may be done in the early morning hours successfully to achieve fruit set.

3.3.7 Delay in fruit ripening

This problem is particularly important in muskmelon and watermelon.

Causal factor

It occurs due to high moisture level and temperature fluctuation at ripening stage.

Symptom

Delay in repening is sometime associated with less sweetness and cracking of fruits.

Management

a) Irrigation should be stopped at the ripening stage to hasten ripening.

b) Sowing time should be adjusted in such a way that fruits ripe in hot and rainless condition which hastens ripening and at the same time improve sweetness of the fruits.

3.3.8 Light belly colour

Symptoms

This disorder is characterized by the undersurface of cucumber fruit remaining light in colour instead of turning dark green.

Causal factor

Commonly occurs on fruits lying on cool, moist soil.

Management

a) Avoid luxuriant vine growth.

3.3.9 Air pollution injury

This injury is seen particularly in watermelon.

Causal factor

This happens due to leaf exposure to ozone or sulpphur di oxide.

Symptoms

Ozone injury appears first on older leaves. Affected leaves appear silvery to whitish and in severe condition leaves are killed.

Management

a) Grow crops away from heavily polluted areas.

3.3.10 Pollination problems

Cucurbits are mostly monoecious plants. There are separate male and female blossoms on the same plant. The male flowers tend to open first followed by the female flowers. The female flower is open for only one day and is most receptive during the morning hours. During this time the flower must receive about 15 bee visits for maximum pollination. Unfertilized or poorly fertilized flowers fall from the vine. Majority of cucurbits start flowering 30-45 days after sowing and it follows a definite sequence. An alternate sequence of male and female flowers follows up to fruit set. The first 4-6 flowering nodes bear male flowers and alter female flowers.

Pollination takes place early in the morning between 6-8 am in cucumber, pumpkin, muskmelon and watermelon. Pollination is altered in the day when temperature is high in bottle gourd and ridge gourd. In snake gourd and pointed gourd, anthesis takes place during night and pollination early in the morning. In pumpkin, pollen production is more while in muskmelon, pollen production is scanty and pollen grains are sticky due to oily film surrounding them. Extent of cross pollination in cucurbits is 60-80%. They are entomophilous and bees, beetles and moths help in pollination. It is only when both the male and female flowers are open pollination can occur. Sex ratio (male: female) of cucurbits flowers ranges from 25-30:1 to 15:1. It is influenced by environmental factors. High N content in the soil, long days and high temperature favour maleness. Besides environmental factors, endogenous levels of auxins, gibberellins, ethylene and abscisic acid also determine sex ratio and sequence of flowering.

The plants may appear to be healthy, growing well, and flowering, but many or all of the blossoms drop from the plant. Fruits may appear to start developing but become incomplete or die entirely. Female flowers have a small fruit under the flower which only develops if the flower is pollinated.

Many factors influence the yield and quality of cucurbit crops, but one important consideration is successful pollination. Poor fruit set and deformed fruit are often the result of inadequate pollination. With insect-pollinated crops one can fertilize the soil, irrigate the crop, control pests and still fail to produce an acceptable crop because of poor pollination. If seeds are not fully developed, poor pollination may be responsible. While adverse temperatures can reduce pollen viability and reduce pollination effectiveness. A more common problem is insufficient bee activity for adequate pollination.

Melon and cucumber flowers are pollinated exclusively by honey bees and other insect pollinators. They are not wind or self-pollinated. Insects are required for pollen transfer because of the large size of the pollen grains, their stickiness, and the way they are released from the anthers.

Also, since these plants typically produce only small amount of pollen, pollinators are needed to efficiently transfer pollen from one flower to the next. While wild bees or distant honeybee colonies may suffice for small melon or cucumber fields, they may be inadequate for the commercial grower whose income depends on substantial yields of high quality fruit. Maximum yield is harvested in a minimum number of trips across the field are economically advantageous. To accomplish this, good pollinator activity is essential.

The ways to meet up the pollination problem

It has been found that for adequate pollination of cucurbits it takes at least nine honey bee visits per flower. Since each bee will visit about 100 flowers per foraging trip, usually at least one strong hive per acre is required. Honey bees are most efficient if they can forage within 200 yards from the hive. Honey bee colonies should be moved into position near the field about the time the first female flowers are seen. If the bees are moved in too early, they may find other attractive flowering plants in the area and not visit the cucurbits.

Hives can be removed from cucurbit fields when flowering begins to diminish or when the vines begin to break down. Leave colonies in cucumber fields until a day or so before the last picking.

Care of Bees

In most of the areas, bees can be rented from a local bee keeper. Optimally, colonies should be located on two or three sides of fields up to 30 acres in size. An ideal location is one that exposes the hives to early morning sun and provides some shade at mid-day.

Only strong colonies should be selected for pollination purposes. A strong colony will contain enough bees to cover 10 to 20 frames and brood in eight or more combs. Bees should be provided with a nearby water source.

Protecting Bees from Pesticide Poisoning

Farmers need to be aware that pesticides used to control insects in the field are also hazardous to bees. Many of the commonly used insecticides remain toxic for up to 24 hours or more after application.

The risk of bee kills can be reduced by applying pesticides late in the evening after bees have completed foraging and by not spraying when dew is heavy. If possible, application of pesticides with lowest possible toxicity to bees to be recommended.

Usually, wettable powders (WP) and dusts are more toxic to bees than emulsifiable formulations. When feasible, plan to control insect pests before moving bees into the fields.

Colour Plates

Red pumpkin beetle infesting pumpkin

Symptoms produced by red pumpkin beetle

Egg mass of Epilachna beetle on bittergourd leaf.

Eplichna beetle grub skeletonising bittergourd leaf.

Adult Epilachna beetle.

Grub of stem boring beetle infesting pointedgourd vine.

Insect pests of cucurbits

Melon fly infestedpumpkin

Adult melon fly

Fruit fly laying eggs on bottlegourd

Pheromone trap

Pheromone trap

Pheromone trap

Insect pests of cucurbits

Summer squash fruit infested by melon fly

Snake gourd semilooper infesting bottle gourd

Stink bug adult

Leaf miner infestation on ash gourd

Adult of pumpkin caterpillar

Pointed gourd vine infested by stem boring beetle

Insect pests of cucurbits

Net blight of pointed gourd

Pythium fruit rot pointed gourd

Collar rot of cucumber

Fusarium wilt on bitter gourd

Phoma blight of cucurbit

Damping off of cucurbit

Diseases of cucurbitaceous vegetables

Downy mildew of cucumber

Powdery mildew of ridge gourd

Alternaria leaf blight of bottle gourd

Angular leaf spot on ridge gourd

Sclerotia rot of pointed gourd

Fruit and vine rot of pointed gourd

Fungal diseases of cucurbits

Spilanthes spp.

Phisalis minima

Borreria spp.

Digitaria sanguinalis

Helitropium indicum

Polygonum persicaria (Knot weed)

Weed pests of cucurbitaceous crops

Cyperus rotandus

Chenopodium album

Solanum nigrum

Portulaca oleracea

Sonchus spp.

Sonchus arvensis

Weed pests of cucurbitaceous crops

Potassium deficiency on pumpkin

Ca deficiency on bottle gourd

Ca deficiency on pumpkin

Magnesium deficiency on pumpkin

Zn deficiency in pumpkin

Boron deficiency

Nutritional disorders in cucurbitaceous crops

Cucumber mosaic virus affected leaf

Green mottle mosaic virus affected leaf

Cucurbit leaf curl virus affected leaf

Sun scorching of cucurbit

Root knot on pointed gourd

Red spider mite

Viral, environmental, nematode and mite infestation on cucurbit

References

Abro, G. H. and Wright, D. J. 1989. Host plant preference and influence of different cabbage cultivars on the toxicity of abamectin and cypermethrin against *Plutella xylostella. Ann. Appl. Biol.* 115: 461-467.

Adam, K. L. 2006. Squash bug and squash vine borer: organic controls. National Sustainable Agriculture Information Service. ATTRA Publication.

Adan, A. P., Estal, D., Budia, F., Gonzalez, M. and Vinuela, E. 1996. Laboratory evaluation of the novel naturally derived compound Spinosad against *Ceratitis capitata. Pestic. Sci.*, 48:261-268.

Adetula, O and Denton, L. 2003. Performance of vegetative and yield accessions of cucumber (*Cucumis sativa* Linn.) Horticultural Society of Nigeria (HORTSON) Proceedings of 21st Annual Conference, 10-13 Nov, 2002.

Aharoni, Y., Copel, A. and Fallik, E. 1993. Hinokitiol (â-thujaplicin), for postharvest decay control on 'Galia' melons. *New Zealand Journal of Crop and Horticultural Science*, 21(2): 165-169.

Ahlawat, Y. S. and Kulshreshtha, U. K. 1977. A new graft transmissible disease of *Sechium edule. Indian Phytopathology*, 30:268-269.

Ahmad, S., Brattsten, L. B., Mullin, C. A. and Yu, S. J. 1986. Enzymes involved in the metabolism of plant allelochemicals. *In:* L. B. Brattsten and S. Ahmad (Eds.), Molecular aspects of insect-plant associations. Plenum, New York, pp-346.

Ahmed, A., Thakral, S. K., Yadav, A. and Balyan, R. S. 2010. Grain yield of wheat as affected by different tillage practices varieties and weed control methods. National Symposium on Integrated Weed Management in the Era of Climate Change, held at NAAS, New Delhi on 21-22 August, pp-19.

Ahmed, M. S., Rasul, M. G., Bashar, M. K., Mian, A. S. M. 2000. Variability and heterosis in snake gourd (*Trichosanthes anguina* L.). *Bangladesh Journal of Plant Breeding and Genetics* 13:27-32.

AICVIP, 1997-98. Progress report, published by ICAR, New Delhi.

Ainsworth, G. C. 1935. Mosaic disease of cucumber. *Annals of Applied Biology*, 22: 55-67.

Akilan,C. A., Srividhya, M., Mohana Priya, C., Jeba Samuel, C. S. and Sundara Mahalingam, M. A. 2014. Comparative analysis of phytochemicals, antibiogram of selected plants in Solanaceae family and its Characterization Studies. *International Journal of Pharmacy and Pharmaceutical Sciences* 6(2):946-950.

Alam, M. Z. 1969. Insect Pests of Vegetables and Their Control in East Pakistan. Agril. Inf. Serv., Dept. Agric., 3, R. K. Mission Road, Dacca. pp-149.

Allwood, A. J., A. Chinajariyawong, R. A. I. Drew, E. L. Hamacek, D. L. Hancock, C. Hengsawad, J. C. Jipanin, M. Jirasurat, C. Kong Krong, S. Kritsaneepaiboon, C. T. S. Leong and S. Vijaysegaran. 1999. Host plant recorded for fruit flies (Diptera: Tephritidae) in Southeast Asia. *Raffles Bull. Zool.* Supplement No.7.: 1-92.

Americanos, P. G. 1991. Control of *Orobanche* in celery. Technical Bulletin-Cyprus Agricultural Research Institute No. 137.

Amin, K. S. and Ullasa, B. A. 1981. Effect of thiophanate on epidemic development of anthracnose and yield of watermelon. *Phytopathology*, 71: 20-22.

Amin, K. S., Ullasa, B. A. and Sohi, H. S. 1982. Quantitative determination of resistance to fungicidal sprayers. *Indian J. Agri. Sci.* 49: 53-57.

Anjaria, J., Parabia, M., Bhatt, G. and Khamar, R. 2002. Natural heals: A glossary of selected indigenous medicinal plants of India, pp 35, Sristi innovations 2nd edition, Ahmedabad.

Anonymous 2004. Susanhata upaye bivinna sashyer rog o poka niyantran (in Eng.: Integrated disease and insect pest management of different crops). Department of Agriculture, Government of West Bengal. pp-34.

Anonymous 2009. Disease identification guide for cucurbits. Syngenta Crop Protection, Inc., P.O. Box 18300, Greensboro, NC 27419. pp-74.

Anonymous 2010. Plant Protection Schedule: Integrated Pest Management. 1st Ed., Derectorate of Agriculture, Govt. of Agriculture, pp. 200.

Anonymous, 1991a. Castor: Annual Report. Directorate of Oilseeds Research, Hyderabad. pp-137.

Anonymous, 1991b. Mites of agricultural importance in India and their management. *Technology Bulletin No.* 1. All India Coordinated Research Project on Agricultural Acarology, UAS, Bangalore, pp-18.

Antignus, Y., Wang, Y., Pearlsman, M., Lachman, O., Lavi, N. and Galon, A. 2001. Biological and molecular characterization of a new cucurbit infecting tobamovirus. *Phytopathology*, 91:565-571.

Arnett, R. H., Thomas, M. C.; Skelley, P. E. and Frank, J. H. 2002. American Beetles. CRC Press, Boca Raton, Florida, pp-861.

Araujo S M and Almedia L. 2004. Behaviour and life cycle of *Epilachna vigintioctopunctata* (Fab.) (Coleoptera: Coccinellidae) in *Lycopersicum esculentum* Mill. (Solanaceae). *Revista Brasileira de Zoologica.* 21(3): 543-550.

Ascard, J., Hansson, D. and Svensson, S. 2014. Physical and cultural weed control in Scandinavia. 10th EWRS Workshop on Physical and Cultural Weed Control, Alnarp, Sweden. Proceedings, pp-2.

Aulakh, K. S. 1971. Damping off of toria seedling due to *Pythium butleri. Indian Phytopath.* 24: 611-612.

Awasthy, P., Bhambri, M. C., Pandey, N., Bajpai, R. K. and Dwivedi, S. K. 2014. Effect of water management and mulches on weed dynamics and yield of maize. *The Ecoscan* 6: 473-478.

Aycock, R. 1966. Stem rot and other diseases caused by *Sclerotium rolfsii* – the status of Rolfs' fungus after 70 years. North Carolina State University Agricultural Experiment Station. *Technical Bulletin No.* 174, pp-202.

Ayyanna, T. and Ramadevi, M. 1987. *Zonabris putulata* Thompson (Order: Coleoptera, Family: Meloidae) and *Chrysocoris purpurea* (Order: Hemiptera, Family: Pentatomidae) as pests of cashew, in Andhra Pradesh. *The Cashew,* 1(3):9-10.

Azam, K.M., 1991. Toxicity of neem oil against leaf miner (*Liriomyza trifolii* Burgess) on cucumber. *Plant Prot. Quart.*6, 196–197. (c.f. R.A.E., A, 1993).

Ba-Angood S. A. S. 2009. Bionomics of the melon worm *Palpita (Diaphania) indica* (Saund) (Pyralidae: Lepidoptera) in PDR of Yemen. *Journal of Applied Entomology,* 88(1-5):332-336.

Back, E.A., and Pemberton, C.E. 1914. *J. Agric. Res.,* 3: 269-274.

Backman, P. A., and Brenneman, T. B. 1984. Compendium of Peanut Diseases. Amer. Phytopath. Soc., St. Paul, Minnesota.

Bag, T. K. 2000. Status of vegetable diseases caused by *Sclerotinia sclerotiorum* in different land use systems of Arunachal Pradesh. *Environment and Ecology,* 18(1): 88-91.

Bains, S. S. and Jhooty, J. S. 1976a. overwintering of *Pseudoperonospora cubensis* causing downy mildew of muskmelon. *Indian Phytopath.* 29:211-213.

Bains, S. S. and Jhooty, J. S. 1976b. Host range and possibility of pathological races in *Pseudoperonospora cubensis,* cause of downy mildew of muskmelon. *Indian Phytopath.* 29:214-216.

Bains, S. S. and Prakash, V. 1985. Susceptibility of different cucurbits to *Pseudoperonospora cubensis* under natural and artificial epiphytotic conditions. *Indian Phytopath.* 38:138-139.

Bains, S. S. and Sharma, N. K. 1986. Two new races of *Pseudoperonospora cubensis* on muskmelon in Punjab. *Indian J. Mycol. Pl. Path.* 7:86.

Bains, S. S., Sokhi, S. S. and Jhooty, J. S. 1977. *Melothria maderaspatana,* a new host of *Pseudoperonospora cubensis. Indian J. Mycol. Pl. Path.* 7: 86.

Baitar, I. E. 2003. Aljamaiul Mufradat-ul Advia Wal Aghzia, p. 248, CCRUM, New Delhi.

Bajwa, D. S. and Mavi, G. S. 1988. Application of low volume concentration with Fogairs for the control of *Aulacophora foveicollis* (Lucas) on muskmelon. *Indian Journal of Entomology,* 47:349-352.

Bandyopadhyay, S. R. and Mukhopadhyay, S. 1977. Incidence of a new disease of muskmelon (*Cucumis melo* L. var. *reticulatus*) in West Bengal. *Curr. Sci.* 46:566-567.

Banerji, R., Sahoo, S. K., Das, S. K. And Jha, S. 2005. Studies on incidence of melon fly, *Bactrocera cucurbitae* (Coq.) in relation to weather parameters on bitter gourd in new alluvial zone of West Bengal, *Journal of Entomological Research,* 29(3):179-182.

Bateman, M. A. 1982. Chemical methods for suppression or eradication of fruit fly populations, In: Drew, R. A. I., Hooper, G. H. S. and Bateman, M. A. (eds.), *Economic Fruit Flies of the South Pacific Region.* 2nd edition, Brisbane, pp-115-128.

Batra H N. 1964 the population behavior, host specificity and development potential of fruit flies bred at constant temperature in winter. *Indian Journal of Entomology,* 20: 195-206.

Behera, T. K. 2004. Heterosis in bittergourd. *Journal of New Seeds* 6: 217-221.

Bem, F. P. and Vassilakos, N. 2000. Occurrence of the disease "Internal deterioration of watermelon fruit" in watermelon cultivation in Greece. *Phytopathologia Mediterranea,* 39:325.

Bhan, M., Bhadauria, U. P. S. and Sharma, S. 2012. Effect of varieties for weed and drought tolerance in different irrigation regimes. In: Proceeding of 3rd International Agronomy Congress, Held at IARI, New Delhi, pp-58-57.

Bharathi, L.K. and K.J. John, 2013. Description and Crop Production in *Momordica* Genus in Asia: An Overview, Springer India, XVII,:5-36.

Bhargava, A. K. and Singh, R. D. 1985. Comparative study of Alternaria blight, losses and causal organisms of cucurbits in Rajasthan. *Indian J. Mycol. Pl. Path.* 15: 150-154.

Bhargava, B. 1977. Some properties of two strains of watermelon mosaic virus. *Phytopath. Z.*, 88: 199-202.

Bhargava, B. and Bhargava, K. S. 1977. Cucurbit mosaic viruses in Gorakhpur. *Indian J. Agri. Sci.*, 47: 1-5.

Bhargava, B., Bhargava, K. S. and Joshi, R. D. 1975. Perpetuation of watermelon mosaic virus in eastern Uttar Pradesh, India. *Pl. Dis. Reptr.*, 59:634-636.

Bhatia, S.K. and Mahto, Y. 1968. Note on breeding of fruit flies *Dacus ciliatus* Loew and *D. cucurbitae* Coquillett in stem galls of *Coccinia indica* W. & A. *Indian Journal of Entomology* 30: 244-245.

Bhattacharya, S. 2000. Koshatak (Jhinga) In: Chiranjeeb Banousadhi – 4th volume (Bengali). Ananda Publishers Pvt. Ltd., Kolkata, India. pp-236-242.

Bhattacharya, S. P., Mondal, B. and Bardolui, S.K. 2006a. Crop pests and its management In: Agricultural Science, Vocational second paper (Class XII). (In Bengali language-Shasya satru-o-tar pratikar. In: Krishi Bidya). West Bengal State Council for Vocational Education and Training, Kolkata, pp-1-118.

Bhattacharya, S. P., Mondal, B. and Bardolui, S.K. 2006b. Shasyosatru o tar protikar (In english: Crop pests and their management). In: Krishi Bidya. West Bengal State Council for Vocational Education and Training, Kolkata, pp-220.

Bhullar, M. S., Kaur, T., Kaur, S. and Yadav, R. 2015. Weed management in vegetable and flower crop-based systems. *Indian Journal of Weed Science* 47(3):277–287.

Blackman, R. L., and V. F. Eastop. 1985. Aphids on the Worid's Crops: An identification guide. Wiley and Sons, Inc., New York.

Bologna, M. A. and Pinto, J. D. 2002. The Old World genera of Meloidae (Coleoptera): a key and synopsis. *J. Natu. Hist.*, 36:2013-2102.

Bose, T. K. and Mitra, S. K. 1990. Fruits: Tropical and Sub-tropical, Naya Prakash, Calcutta, India. pp-838.

Brodie B. B., Evans, K. and Franco, J. 1993. Nematode parasites of potatoes. *In*: Evans K, Trudgill D. L., Webster J. M., eds. *Plant Parasitic Nematodes in Temperate Agriculture*. Wallingford, UK: CAB International, 87–132.

Brust, G. E. 2010. Squash vine borer (Lepidoptera: Sesiidae) management in pumpkin in the mid-Atlantic. *Journal of Applied Entomology* 134: 781-788.

Butani, D. K. 1975. Insect pests of fruit crops and their control-15: Date palm. *Pesticides*, 9(3):40-42.

Butani, D. K. and Jotwani, M. G. 1984. Insects in Vegetables. Periodical Expert Book Agency, New Delhi, pp-356.

Butani, D. K., Jotwani, M. J. and Verma, S. 1977. Insect pests of vegetables and their control-Cole crops. *Pesticides,* 11(5):19-24.

Butler, E. J. and Bisby, G. R. 1931. The fungi of India. Sci. Monograph I. ICAR, New Delhi.

Canhilal, R., G. R. Carner, R. P Griffin, D. M. Jackson, and D. R Alvarez. 2006. Life history of the squash vine borer, *Melittia cucurbitae* (Harris) (Lepidoptera: Sesiidae) in South Carolina. *The Journal of Agricultural and Urban Entomology,* 23:1-7.

Capinera, J.L. 2001. Handbook of Vegetable Pests. Academic Press, San Diego. 729 pp.

Capoor, S. P. and Ahmad, R. J. 1975. Yellow vein mosaic disease of field pumpkin and its relationship with the vector *Bemisia tabaci. Indian Phytopath.* 28:41.

Capoor, S. P. and Verma, L. M. 1948. A mosaic disease of *Lagenaria vulgaris* Ser. in the Bombay Province. *Curr. Sci.* 17:274-275.

Chadha, K. L. 2001. Handbook of Horticulture. Directorate of Information and Publications of Agriculture. ICAR, Krishi Anusandhan Bhavan, Pusa, New Delhi, pp-1031.

Chahal, D. S., Chohan, J. S. and Sidhu, G. S. 1970. Alternaria leaf spot of cucurbits in Punjab. *Indian Phytopathology,* 23:580-581.

Chakravorty, D. P., Ghosh, G. P. and Dhura, S. P. 1969. Repellent properties of thionimone on red pumpkin beetle, *Aulacophora foveicollis* (Lucas), *Technology,* 6:48-49.

Chan, C. K., Forbes, A. R.and Raworth, D. A. 1991. Aphid transmitted viruses and their vectors of the world. Agric. *Canada Res. Branch Technical Bulletin,* 1991-3E:216.

Chand, P. and Singh, A. P. 1976. Aulacophora foveicollis (Lucas) as a pest of Lucerne at Ranchi, Bihar. *Entomologists' Newsletter,* Ranchi Agricultural College, Bihar, India, 6(3):32.

Chandranath, H. T. and Katti, P. 2010. Management of epilachna beetle on ashwagandha. *Karnataka Journal of Agricultural Sciences,* 23(1):171.

Chatterjee, R. and Maitra, S. 2014. An Inventory of the diversity and ethnomedicinal properties of cucurbitaceous vegetables in the Homestead Gardens of Sub Himalayan Districts of West Bengal, India. Research and Reviews: *Journal of Botanical Sciences* 3(4):40-45.

Chattopadhyay, S. B. and Mustafee, T. P. 2008. Plant Diseases and their Management. Aditya Books Private Limited, New Delhi, pp-627.

Chattopadhyay, S. B. and Mustafee, T. P. 1967. Bulletin of Botanical Society of Bengal 21:103-106.

Chattopadhyay, S. B. and Mustafee, T. P. 2008. Plant diseases and their management. Aditya Books Private Limited, New Delhi, pp-627.

Chattopadhyay, S. B. and Sengupta, S. K. 1951. *Bulletin of Botanical Society of Bengal* 5: 58-61.

Chaudhary, F. K. and Patel, G. M. 2007. Biology of melon fly, *Bactrocera cucurbitae* (Coq.) on pumpkin. *Indian Journal of Entomology*, 69(2): 168-171.

Chaudhuri, S. 1975. Fruit rot of Trichosanthes dioica L. caused by Pythium cucurbitacearum Takimoto in West Bengal. *Current Science*, 44(2): 68.

Chaurasia, S. C. 1980. Studies on the foot rot and leaf rot diseases of Pan (*Piper* betle). IX. The effect of various substances on the oxygen uptake of *Phytophthora parasitica* var. *piperina*. *Journal of Phytopathology*, 98(2): 163-170.

Chawla, S.S. 1996. Some critical observations on the biology of melon fly, *Dacus cucurbitae* Coquillet (Diptera: Tephritidae). *Research Bulletin of Punjab University*, 17:105-109.

Cheema, S. S., Kang, S. S. and Bansal, R. D. 1999. Viral diseases of cucurbitaceous crops and their management. In: Diseases of horticultural crops-vegetables, ornamentals and musrooms (Verma and Sharma eds.), Indus Publishing Co., New Delhi, 735pp.

Chelliah, S. 1970. Host influence on the development of melon fly, *Dacus cucurbitae* Coquillet. *Indian Journal of Entomology*, 32:381-383.

Chen, H. Y., Zhao, W. J., Gu, Q. S., Chen, Q., Lin, S. M. and Zhu, S. F. 2008. Real time TaqMan RT-PCR assay for the detection of *Cucumber green mottle mosaic virus*. *Journal of Virological Methods*, 149:326-329.

Chopra, B. L. and Jhooty, J. S. 1974. Biochemical changes in a resistant and susceptible variety of watermelon due to infection by *Alternaria cucumerina*. *Indian Phytopath.*, 27:502-507.

Choudhary, B. R. and Fageria, M. S. 2002. Breeding for multiple disease resistance in cucurbits (water melon, musk melon, cucumber and squash)- a review. *Agnc. Rev.*, 23(4):300-304.

Choudhury, B. 1990. Vegetables. 8th Ed., National Book Trust, New Delhi, pp-195.

Chowdhury, A. K., De, B. K. and Raj, S. K. 1998. Present status of vegetable diseases and suggested future lines of research work in West Bengal. In: Proceedings of National seminar on vegetable diseases and their management strategies in West Bengal (Sengupta, P. K. and Chowdhury, A. K. eds.). Department of Plant Pathology, Bidhan Chandra Krishi Viswavidyalaya. pp-1-9.

Chughtai, C. G. and Baloch, V. K. 1988. Insecticidal control of melon fruit fly. *Pakistan Journal of Entomological Research,* 9:192-194.

Cirulli, M. and Ciccarese, F. 1981. Effect of mineral fertilizers on the incidence of blossom- end rot of watermelon. *Phytopathology,* 71:50-53.

Clark, C. A., and Moyer, J. W. 1988. Compendium of Sweet Potato Diseases. Amer. Phytopath. Soc., St. Paul, Minnesota.

Clausen, C. P., Clancy, D. W. and Chock, Q.C. 1965. Biological control of the Oriental fruit fly (*Dacus dorsalis* Hendel.) and other fruit flies in Hawaii. Technical Bulletin, United States Department of Agriculture, 1322:1-102.

Cogan, B. H. and Munro, H. K. 1980. Family Tephritidae, In: R. W. (eds.), *Catalogue of the Diptera of Afrotropical Region.* British Museum (Natural History), London. pp-518-554. Dhillon, M. K., Singh, R.; Naresh, J. S. and Sharma, H. C. 2005. The melon fruit fly, *Bactrocera cucurbitae*: A review of its biology and management. *Journal of Insect Science,* 5:40.

Connolly, B. 2005. Saving cucurbit seed. *The Natural Farmer,* 2(65):13–15.

Dalela, G. G. 1956. A species of *Rhizopus* causing drying of young fruits of *Cucurbita maxima* L. *Indian Phytopathology,* 9:197.

Darekar, R. N. and Sawant, D. M. 1989. Relationship of bottle gourd mosaic virus with its aphid vectors. Proc. 41st Ann. Meet. *Indian Phytopath. Soc.,* 42: 340.

Das, D. 2015. Cultural attributes of Plants of Darjeeling with potential use and threat through Environmental Degradation in Eastern Himalaya. *Indian Journal of Applied and Pure Biology* 30(1):41-53.

Das, T. K. 2008. Weed science: basics and application.pp 901 Jain Brothers Pub, New Delhi.

Das, T. K., Tuti, M. D., Sharma, R., Paul, T. and Mirja, P. R. 2012. Weed management research in India: An overview. *Indian Journal of Agronomy* 57(3):148-156.

Dasgupta, M. K and Mandal, N. C. 1988. Ecology and Epidemiology of post Harvest disease of perishable: Prelude to their control. *Review of Tropical Plant Pathology,* 2 (1985/86). Today and Tomorrow's Printers and Publishers, New Delhi.

Dasgupta, M. K. 1998. *Meloidogyne* – at the acme of nematode parasitism with special reference to vegetables. In: Proceedings of National seminar on vegetable diseases and their management strategies in West Bengal (Sengupta, P. K. and Chowdhury, A. K. eds.). Department of Plant Pathology, Bidhan Chandra Krishi Viswavidyalaya. pp-39-47.

Dasgupta, M. K. and Mandal, N. C. 1988. Post Hawest Diseases of Perishables. Oxford and IBH Publishing Co. Pvt. Ltd., Calcutta.

Dasgupta, M. K. and Mandal, N. C. 1989. Postharvest Pathology of Perishables. Oxford and IBH Publishing Co. Ltd., New Delhi, pp-623.

De, N., Singh, K. P. and Rai, M. 2003. Micronutrient deficiency symptoms in vegetable crops and their management. *Technical Bulletin No. 14. IIVR* (Varanasi), pp-28.

Deng, Y. L., Li, Z. Y. and Zhang, H. R. 2006. Population dynamics of B. dorsalis, B. cucurbitae and B. tau (Diptera: Tephritidae) in Xishuangbanna. *South-West China Journal of Agricultural Sciences*, 19(4):643-648.

Dhamdhere, S. V., Koshta, V. K. and Rawat, R. R. 1990. Effect of food plants on the biology of *Henosepilachna vigintioctopunctata* (Fab.) (Coleoptera: Coccinellidae). *Journal of Entomological Research*, 14(2):142-145.

Dhillon, M. K., Singh, R.; Naresh, J. S. and Sharma, H. C. 2005a. Reaction of different bitter gourd (*Momordica charantia* L.) genotypes to melon fly, *Bactrocera cucurbitae* (Coq.). *Indian Journal of Plant Protection*, 33(1): 55-59.

Dhillon, M. K., Singh, R.; Naresh, J. S. and Sharma, H. C. 2005b. The melon fruit fly, *Bactrocera cucurbitae*: A review of its biology and management. *Journal of Insect Science*, 5:40.

Dixon, A. F. G. 1987. Evolution and adaptive significance of cyclical parthenogenesis in aphids, *In A.* K. Minks and P. Harrewijn [eds], Worid Crop Pests: Aphids, their biology, natural enemies and control, vol. 2A. Elsevier, Amsterdam, The Netherlands. pp. 289-297.

Doharey, K. L. 1983. Efficacy of some insecticides against fruit flies. *Indian Jouranl of Entomology*, 45(4): 465-469.

Doymaz, I. 2007. The kinetics of forced convective air-drying of pumpkin slices. *Journal of Food Engineering* 79: 243–248.

Drew, R. A. I. 1982. Fruit fly collecting, In: Drew, R. A. I., Hooper, G. H. S. and Bateman, M. A. (eds.), *Economc Fruit Flies of the South Pacific Region*. 2nd ed. Brisbane. Pp-129-139.

Drew, R.A.I. 1989. The tropical fruit flies (Diptera: Tephritidae: Dacinae) of the Australasian and Oceanian Regions. Memoirs of the Queensland Museum 26: 1–521.Dubey, G. S., Nariani, T. K. and Prakash, N. (1974). Identification of snake gourd mosaic virus. *Indian Phytopath*. 27: 470-474.

Dutta, H. L. 1943. *Indian Journal of Agricultural Sciences*, 13:1-17.

Dutta, M. and Singh, B. V. 1989. Blister beetle (*Mylabris phalerata*), a serious pest of pigeonpea in the lower hills of Uttar Pradesh. *International Pigeonpea Newsletter*, 5 (10):29-32.

Eta CR. 1985 Eradication of the melon fly from Shortland Islands (special report). *Solomon Islands Agricultural Quarantine Service, Annual Report*. Ministry of Agriculture and Lands, Honiara.

Fang, M. N. 1989. A nonpesticide method for the control of melon fly, *D. cucurbitae* Coq. *Special Publication of the Taichung District Agricultural Improvement Station,* 16: 193-205.

Felt, F. P. 1919. "New Philippine Gall Midges", *Philipp. J. Sci.,* 14:287-94.

Felton, G. W. and Duffey, S. S. 1990. Inactivation of baculovirus by quinones formed in insect-damaged plant tissues. J. *Chem. Ecol.* 16: 1221-1236.

Felton, G. W., Duffey, S. S., Vail, P. V., Kaya, H. K. and Manning, J. 1987. Interaction of nuclear polyhedrosis virus with catechols: potential incompatibility of host-plant resistance against noctuid larvae. *J. Chem. Ecol.* 13: 947-957.

Fernendo, H. F. 1957. The biology and control of Leptoglossus membranaceous Fab. (Coreidae: Hemiptera). *Tropical Agriculturist,* 113(2):107-118.

Ferreira, S. A. and Boley, R. A. 1992. *Colletotrichum lagenarium.* Crop Knowledge Master. Department of Plant Pathology, CTAHR. University of Hawaii (Manoa).

Fletcher, T. B. 1920. Life histories of Indian insects Micro Lepidoptera: I-Pterophoridae. *Memoirs of Department of Agriculture in India, Entomological Series,* 6(1):9-13.

Franski, R. I. B. and Hatta, T. 1980. Cucumber mosaic virus – variation and problems of identification. *Acta Hort.,* 11:167-174.

Francki, R. I. B., Hu, J. and Palukaltis, P. 1986. Taxonomy of cucurbit infecting tobamovirus as determined by serological and molecular hybridization analyses. *Intervirology,* 26: 156-163.

Froggatt, W.W. 1923. Insect pests of the cultivated cotton plant. No. 4. Cutworms and leaf-eating beetles. *Agric. Gaz. N.S.W.* 34:343-348.

Fu, C., Shi, H. and Li, Q. A. 2006. Review on pharmacological activities and utilization technologies of Pumpkin. *Plant Foods and Human Nutrition,* 61:73–80.

Fukuda, M., Meshi, T., Okada, Y., Otsuki, Y. and Takebe, I. 1981. Correlation between particle multiplicity and location of virion RNA of the assembly initiation site for viruses of the tobacco mosaic virus group. *Proc. Nat. Acad. Sci. USA.* 78:4231-4235.

Galande, S.M., Mote, U.N. and Ghorpade, S. A. 2004. New host plants of serpentine leaf miner, *Liriomyza trifolii* in western Maharashtra. *Ann. Pl. Prot. Sci.,* 12:425-475.

Ganapathy, N., Durairaj C. and Karuppuchamy 2010. Bio- ecology and management of serpentine leaf miner, *Liriomyza trifolii* (Burgess) in cowpea, *Karnataka Journsal of Agricultural Sciences,* 23(1):9-10.

Gangopadhyay, S. and Sharma, R. K. 1976. Spongy rot of pumpkin. *Indian Phytopathology,* 29:423-424.

Gautam, S., Anju Meshram and Srivastava, N. 2014. A brief study on phytochemical compounds present in *Coccinia cordifolia* for their medicinal, pharmacological and industrial applications. *World Journal of Pharmacy and Pharmaceutical Sciences* 3 (2):1995-2016.

Ghosh, S. K. and Mukhopadhyay, S. 1979. Viruses of pumpkin (*Cucurbita moschata*) in West Bengal. *Phytopath. Z.*, 94:172-184.

Ghosh, S. K. and Mukhopadhyay, S. 1979b. Incidence of squirting cucumber mosaic virus of cucumber in West Bengal. *Current Science*, 48(7):788.

Ghosh, S.K. and Senapati, S.K. 2001. Biology and seasonal fluctuation of *Henosepilachna vigintioctoctopunctata* Fabr. On brinjal under terai region of West Bengal. *Indian Journal of Agricultural Research*, 35: 149-154.

Giri, A. N., Deshmukh, M. N. and Gore, S. B. 2006. Effect of cultural and integrated methods of weed control on cotton, intercrop yield and weed control efficiency in cotton based cropping system. *Indian Journal of Agronomy* 51(1):34-36.

Giri, B. K. 1985. Studies on tomato mosaic. Ph. D. Thesis. PG School, IARI, New Delhi.

Giri, B. K., Chenulu, V. V. and Mishra, M. D. 1981. A new graft-transmissible disease of *Momordica chrantia* L. Discussion on Current Problems on Virus and Virus-like Plant Diseases, Kalimpong. Oct. 19-21. IARI Reg. Station, Kalimpong.

Gould, F., Kennedy, G. G. and Johnson, M. T. 1991. Effects of natural enemies on the rate of herbivore adaptation to resistant host plants. *Entomologia Experimentalis et Applicata*, 31:175-180.

Gour, T. B., Ramesh Babu, T., Sriramulu, M., Reddy, D. D. R., Chandrasekhara Rao, K., Ravindra Babu, R. and Narayan Reddy, P. 2008. Crucifers, cucurbits: insect pests, diseases, nutritional disorders. Agricultural Information and Communication Centre, Acharya N. G. Ranga Agricultural University, Rajendranagar, Hyderabad, pp. 88.

Grabowski, M. 2013. Common diseases of cucurbit crops in Minnesota, University of Minnesota, http://www.extension.umn.edu/distribution/horticulture/DG1172-1.html. Last up-dated on 13th July, 2014.

Grewal, J. S. and Malhi, C. S. 1987. *Prunus persica* Batsch damage by birds and fruit fly pests in Ludhiana (Punjab). *Journal of Entomological Research*, 11: 119-120.

Guharoy, S., Bhattacharyya, S., Mukherjee, S., Mandal, N. and Khatua, D. C. 2006. *Phytophthora melonis* associated with Fruit and vine rot disease of pointed gourd in India as revealed by RFLP and sequencing of ITS region. *Journal of Phytopathology*, 154: 612-615.

Guine, R. P. F., Henrriques, F. and Barroca, M. J. 2012. Mass transfer coefficients for the drying of Pumpkin (*Cucurbita moschata*) and dried product quality. *Food Bioprocess Technology* 5:176–183.

Gupta, A. K., Sharma, R. C. and Sharma, S. 2001. Fungal diseases of cucurbits. In: Gupta,V. K. and Paul, Y. S. (eds.). Diseases of vegetable crops. Kalyani Publishers, New Delhi, pp-71-86.

Gupta, D. and Bhatia, R. 2000. Population fluctuation of *Bactrocera* spp. in sub-mountainous mango and guava orchards, *Journal of Applied Horticulture,* 1(2):101-102.

Gupta, D., Verma, A. K. 1995. Host specific demographic studies of the melon fruit fly, *Dacus cucurbitae* Coquillett (Diptera: Tephritidae). *Journal of Insect Science,* 8:87-89.

Gupta, J.N and Verma, A.N. 1978. Screening of different cucurbit crops for the attack of the melon fruit fly, *Dacus cucurbitae* Coq. (Diptera: Tephritidae) *Haryana Journal of Horticulture Science.* 1978;7:78–82.

Gupta, S. K. 1991. The mites of agricultural importance in India with remarks on their economic status. In: *Modern Acarology,* (Eds. Dushabek, F. and Bukva, U.), Academic Press, Hague, pp-509-522.

Halder, J., Rai, A.B., Dey, D. and Kodandaram, M. H. 2014. Is *Apanteles paludicole* Cameron, a Limiting Biotic Factor for Minor Pest Status of Sphenarches caf er (Zeller)? *Journal of Biological Control,* 28(2):119-121.

Hall, D. H., Gubler, W. D. and Sciaroni, R. H. 1981. Fusarium fruit rot of cucurbits. *Calif. Plant Pathol.,* 54:3.

Hall, M. J. R. 1984. Trap-oriented behaviour of red-banded blister beetles, *Mylabris designata* var. *hacolyssa* Rochebrune (Coleoptera: Meloidae), in the Sudan. *Bull. Entomol. Res.,* 74(01):103-112.

Hancock, D. L. 1981. Some economic Zimbabwean fruit flies (Diptera: Tephritidae). *Hortus.* (Zimbabwe), 27:11-15.

Hancock, D.L. 1989. Pest status; southern Africa. In: World crop pests 3(A). Fruit flies; their biology, natural enemies and control (Ed. by Robinson, A.S.; Hooper, G.), pp. 51-58. Elsevier, Amsterdam, Netherlands.

Hara, A. H., Kaya, H. K., Gaugler, R., Lebeck, L. M. and Mello, C. L. 1993. Entomopathogenic nematodes for biological control of the leafminer, *Liriomyza trifolii* (Dipt.: Agromyzidae). *Entomophaga,* 38:359-369.

Hardy, D. E. 1973. Fruit flies (Diptera: Tephritidae) of Thailand and bordering countries. *Pacific Insects Monograph,* 31:340-353.

Hare, J. D. 1992. Effects of plant variation on herbivore-natural enemy interactions. *In:* R. S. Fritz and E. L. Simms (Eds.), Plant resistance to herbivores and pathogens, Univ. of Chicago Press, Chicago, pp-590.

Hare, J. D. and Andreadis, T. G. 1983. Variation in the susceptibility of *Leptinotarsa decemliniata* (Coleoptera: Chrysomelidae) when reared on different host plants to the fungal pathogen *Beauveria bassiana* in the field and laboratory. *Environ. Entomol*. 12: 1892-1897.

Hasyim, A., Murryati and de-Kogel, W. J. 2008. Population fluctuation of adult males of the fruit fly, *Bactrocera tau* Walker (Diptera: Tephritidae) in passion fruit orchards in relation to abiotic factors and sanitation. *Indonesian Journal of Agricultural Science*, 9(1):29-33.

Hazra, P. and Som, M. G. 2015. Vegetable Science..Kalyani Publishers, New Delhi. pp-150-163.

Henneberry, T. J., L. Forlow Jech, T. de la Torre, and D. L. Hendrix 2000. Cotton aphid (Homoptera: Aphididae) biology, honeydew production, sugar quality and quantity, and relationships to sticky cotton. *Southwest. Entomol.* 25: 161-174.

Hill, D. S. 1975. Agricultural insect pests of the tropics and their control. Cambridge University Press, Cambridge. pp-516.

Hiremath, P. C. and Govindu, H. C. 1973. Studies on market diseases of vegetables in Mysore State. I. [*Alternaria tenuis, Rhizopus nigricans, Curvularia lunata*]. *Mysore Journal of Agricultural Sciences*, 7 (3): 428-435.

Hodges, D. M. 2003 Postharvest Oxidative Stress in Horticultural Crops, Food Products Press, New York, USA., pp-11-12.

Hoffman, M. P. and Zitter, T. A. 1994. Cucumber beetles, corn rootworms, and bacterial wilt in cucurbits. Fact sheet. Department of Plant Pathology Cornell University, Ithaca, New York, p. 781.00.

Hollingsworth, R., Vagalo, M., Tsatsia, F. 1997. Biology of melon fly, with special reference to the Solomon Islands. In: (Edited by Allwood A.J. and Drew R.A.I.) Management of fruit flies in the Pacific. *Proceedings of Australian Country Industrial Agricultural Research*, 76: 140-144.

Hossain, M. S., Khan, A. B. Haque, M. A., Mannan, M. A. and Dash, C. K. 2009. Effect of different host plants on growth and development of epilachna beetle. *Bangladesh Journal of Agricultural Research*, 34(3):403-410.

Hsieh, W. H. and Goh, T. K. 1990. Cercospora and similar fungi from Taiwan. Maw Chang Book Company, Taipei.

Hutson, J. C. 1936. Entomological Notes. *Tropical Agriculturist*, 87(5):289-295. In: The Viruses: The Plant Viruses (Milna, R. G. ed.). Plenum Press, New York, pp- 367-378.

Inayatullah, C., Khan, L. And Manzoor-Ul, H. 1991. Relationship between fruit infestation and the density of melon fruit fly adults and puparia. *Indian Journal of Entomology*, 53(2):239-243.

Inderjit, Weston, L. A. and Duke, S. O. 2005. Challenge, achievements and opportunities in allelopathy research. *Journal of Plant Interactions*, 1(2): 69-81.

Inomata, S. I., Komoda, M., Watanabe, H., Nomura, M. and Ando, T. 2000. Identification of sex pheromones of *Anadevidia peponis* and *Macdunnoughia confusa* and field tests of their role in reproductive isolation of closely related Plusiinae moths. *Journal of Chemical Ecology*, 26(2):443-455.

Islam, K., Islam, M.S. and Ferdousi, Z. 2011. Control of *Epilachna vigintioctopunctata* Fab. (Coleoptera: Coccinellidae) using some indigenous plant extracts. *Journal of Life and Earth Sciences*, 6: 75-80.

Islam, N., Baltazar, A. M., De Datta, S. K. and Karim, A. N. M. R. 2009. Management of Purple Nutsedge (*Cyperus rotundus* L.) Tuber Populations in Rice-Onion Cropping Systems. *Philippine Agriculture Scientist*, 92 (4): 407-418.

Jackson, C.G., Long, J.P and Klungness, L.M. 1998. Depth of pupation in four species of fruit flies (Diptera: Tephritidae) in sand with and without moisture. *Journal of Economic Entomology*. 91: 138–142.

Jadhav, L. D., Kadam, M. V.; Ajri, D. S.; and Pokharkar, R. N. 1979. Host preference of leaf footed plant bug, *Leptoglossus membranaceous* Fab. *Indian Journal of Entomology*, 41(3):299-300.

Jaya Kumar, R. and Jagannathan, R. 2003. Weed Science Principles,pp 338. Kalyani Publishers, New Delhi.

Jeyakumar, P. 1995. Studies on Biology and management of serpentine leaf miner, *Liriomyza trifolii* (Burgess) (Diptera:Agromyzidae) on cotton. *M.Sc (Ag.) Thesis.* Tamil Nadu Agric. Univ., Coimbatore, pp-145.

Jhala, R. C., Bharpoda, T. M.; Charda, A. J. and Sisodiya, D. B. 2004. *Leptoglossus australis* Fab. (Heteroptera: Coreidae) on bitter gourd in Gujrat. *Insect Environment*, 9(4):177-178.

Jhooty, J. S. 1967. Identity of powdery mildew of cucurbits in India. *Pl. Dis. Reptr.* 51:1079-1080.

Jhooty, J. S. and Grover, R. K. 1971. Rhizoctonia root rot of cucurbits and its control in India. *Ind. Phytopath.* 24: 571-574.

Jhooty, J. S. and Singh, R. S. 1971. Charcoal rot of melon – a new record for India. *Indian Phytopathology* 24: 578-579.

Johri, R. and Johri, P. K. 2003. Food preferences of red pumpkin beetle, *Aulacophora foveicollis (Lucas) at Kanpur in Uttar Pradesh, India. Journal of Applied Zoological Researches*, 14(1):80-81.

Joseph, P. J. and Ramanathan, M. M. 1978. Studies on the mosaic disease of snake gourd (*Trichosanthes anguina* L.). *Agri. Res. J. Kerala*, 16: 148-154.

Joshi, R. D. 1977. Efficacy of *Aphis gossypii* as vector of turnip mosaic virus and water melon mosaic virus. *Indian Phytopath.*, 30:541-544.

Kannaiyan, S. and Prasad, N. N. 1975. Saprophytic activity of muskmelon wilt fungus in soil. *Indian Phytopath.*, 28:167-170.

Kapoor, V. C. 1970. Taxonomy and Biology of Economically Important Fruit Flies of India. *Isr. J. Entomol.* 35-36(6):459-475.

Kapoor, V.C. and Agarwal, M.L. 1983. Fruit flies and their natural enemies in India,104-105. In Cavalloro, R. (ed.). Proceedings of the CEC/IOBC International symposium on Fruit flies of economic importance, Athens, Greece, 1982. Balkema Publications, Rotterdam.

Kaur, J. and Jhooty, J. S. 1985. Presence of race 3 of *Sphaerotheca fuliginea* on muskmelon in Punjab. *Indian Phytopath.*, 39:297-299.

Kaur, J. and Jhooty, J. S. 1986. Pathological specialization in *Sphaerotheca fuliginea* causing powdery mildew of cucurbits. *Indian Phytopath.*, 38:302-305.

Kaushal, K. K. 1999. Proceeding of National symposium on rational approaches in nematode management for sustainable agriculture, Anand, India, pp-16-19.

Kawate, M. K.; Coughlin, J. A. 1995. Increased green onion yields associated with abamectin treatments for *Liriomyza sativae* (Diptera: Agromyzidae) and *Thrips tabaci* (Thysanoptera: Thripidae). *Proceedings of the Hawaiian Entomological Society* 32:103-112.

Keeley, P. E. 1987. Interference and interaction of purple Nutsedge and yellow Nutsedge with crops. *Weed Technology*, 1:74-81.

Kennelly, M. 2012. Diseases of cucurbits crops. Proceedings of Great Plains Growers Conference held on 6th Jan, 2012, at Kansas State University, Department of Plant Pathology, pp-1-20.

Kesavan, R. and Prasad, N. N. 1974. Correlation between crude cucurbitacin content in certain muskmelon varieties and *Fusarium* wilt incidence. *Indian J. Exp. Biol.* 12:476-477.

Kewat, M. L. 2014. Improved weed management in Rabi crops. Proceedings of National Training on Advances in Weed Management pp-22-25.

Khan, A. M., Khan, M. W., Akram, M. and Khan, A. H. 1976. Studies on the cucurbit powdery mildew. II. Varietal response of some cucurbits to *Sphaerotheca fuliginea*. *Indian Phytopath.* 29:306-308.

Khan, L. 1987. Biology and control of melon fruit fly, *Dacus cucurbitae* (Coq.) Ph. D. thesis, University of Agricultural Sciences, Faisalabad, Pakistan, pp-43.

Khan, L., Haq, M.U., Mohsin, A.U and Inayat-Tullah, C. 1993. Biology and behavior of melon fruit fly, Dacus cucurbitae Coq. (Diptera: Tephritidae) *Pakistan Journal of Zoology*. 25:203–208.

Khan, M. M. H. 2012. Morphometrics of cucurbit longicorn (*Apomecyna saltator* F.) coleoptera: cerambycidae reared on cucurbit vines. *Bangladesh J. Agril. Res.*, 37(3):543-546.

Khan, M. W., Khan, A. M. and Akram, M. 1972. Perithecial stage of certain powdery mildews including some new records. *Indian Phytopath.* 25: 220-224.

Khan, M.A., Hussain, N and Sattar, S, 2011. Response of *Myzus persicae* (Sulzer) to Imidacloprid and Thiamethoxam on susceptible and resistant potato varieties. *Sarhad J. Agric*, Vol.27, No.2, pp-263-269.

Khan, S. M and Wasim, M. 2001. Assessment of different plant extracts for their repellency against red pumpkin beetle, *Aulacophora foveicollis* (Lucas) attacking muskmelon (*Cucumis melo* L.) crop. *Journal of Biological Sciences*, 1(4):198-200.

Khan, S. M., Majid, A. and Habib, M. 1985. Insect pest of muskmelon and their control. Zaraat Nama, Lahore, 1st March, 1985.

Khandelwal, G. L. and Prasad, R. 1979. Nutritional requirements of *Alternaria cucumerina* causing blight disease of cucurbits. *Indian J. Mycol. Pl. Path.* 9: 237-241.

Khatua, D. C. and Maiti, S. 1982. Two new leaf blight diseases caused by *Rhizoctonia solani. Indian Phytopathology* 35:124-125.

Khatua, D. C. and Saha, G. 2004. Diseases of pointed gourd. In: Plant Pathology: Problems and Prospectives. Bidhan Chandra Krishi Viswavidyalaya, Mohanpur, pp-101-109.

Khatua, D. C., Chakravarty, D. K. and Sen, C. 1981a. *Phytophthora cinnamomi* causing stem and fruit of *Trichosanthes dioica. Indian Phytopathology,* 34: 373-374.

Khatua, D. C., Das, A., Ghanti, P. and Sen, C. 1981b. Downy mildew of cucurbits in West Bengal and preliminary field assessment of fungicides against downy mildew of cucumber. *Pestology,* 5(2):30-31.

Khatua, D. C., Mondal, B. and Saha, G. 2013. Bioassay of some agricultural chemicals, human drugs and disinfectants on *Phytophthora melonis* Katsura causing fruit and vine rot of pointed gourd. *The Journal of Plant Protection Sciences,* 5(1):32-36.

Khatua, D. C., Mondal, B., Hansda, S. and Ray, S. K. 2014. Sclerotinia rot of ridge gourd and pointed gourd in lateritic zone of West Bengal. *Scholar Academic Journal of Biosciences,* 2(4):251-254.

Khorsheduzzaman, A.K.M., Z. Nessa and M.A. Rahman, 2010. Evaluation of mosquito net barrier on cucurbit seedling with other chemical, mechanical and botanical approaches for suppression of red pumpkin beetle damage in cucurbit. *Bangladesh Journal of Agricultural Research*, 35(3):395-401.

Khosla, H. K., Dave, G. S. and Nema, K. G. 1973. Identity of powdery mildew on white gourd (*Benincasa hispida* cogn.) in Madhya Pradesh. *JNKVV Res. J.* 7:175-177.

Khulakpam, N. S., Singh, V. and Rana, D. K. 2015. Medicinal Importance of Cucurbitaceous Crops. *International Research Journal of Biological Sciences*, 4(6):1-3.

King, J. R. and Hennessey, M. K. 1996. Spinosad bait for the Caibbean fruit fly (Diptera: Tephritidae). *Fla. Entomol.*, 79:526-531.

Klahn, S. A. 1987. Cantharidin in the natural history of the Meloidae (Coleoptera). Colorado State University, pp-272.

Klungness, L. M., Jang, E. B., Man, R. F. L., Vargas, R. I., Sugano, J. S. and Eujitani, E. 2005. New sanitation techniques for controlling Tephritid fruit flies (Diptera: Tephritidae) in Hawaii. *Journal of Applied Sciences and Environmental Management*, 9(2):5-14.

Komuro, Y. 1971. Cumber green mottle mosaic virus on cucumber and watermelon and melon necrotic spot virus on muskmelon. *Japan Agricultural Research Quarterly*, 6: 41-45.

Komuro, Y., Tochihara, H., Fukatsu, R., Nagani, Y. and Yoneyama, S. 1968. Cucumber green mottle mosaic virus on watermelon in Chiba and Ibaraki prefectures. *Annals of Phytopathological Society of Japan*, 34:377.

Konar, A., Mohasin, M. 2002. Incidence of Epilachna beetle at different localities of west Bengal. *J. Indian Potato Assoc.* 29(1-2):95-97.

Konar, A., Roy, P. S. and Paul, S. 2005. Bioefficacy of insecticides and biopesticides against *Henosepilachna vigintioctopunctata* Fab. (Coccinellidae: Coleoptera) on potato. *Journal of Interacademicia*, 9(2):268-270.

Kore, S. S. and Kharwade, L. R. 1987. Fruit rot of round gourd caused by *Fusarium oxysporum. Indian Phytopathology* 40:251.

Koul, V. K., Bhagat, K. C. 1994. Biology of melon fruit fly, *Bactrocera* (*Dacus*) *cucurbitae* Coquillett (Diptera: Tephritidae) on bottle gourd. *Pest Management and Economic Zoology*, 2:123-125.

Krawinkel, M. B. and Keding, G. B. 2014. Bitter Gourd (*Momordica charantia*): A dietary approach to hyperglycemia. *Nutrition reviews*, 64(7):331-337.

Krishnamurthy, V., Goswami, B. K. and Gupta, J. N. 1972. *Helminthosporium rostratum* causing a leaf spot of cucurbitaceous host muskmelon (*Cucumis melo*). *Indian Phytopath.* 25:302-303.

Kulkarni, S. A. and Kulkarni, S. 1994. Biological control of *Sclerotium rolfsii* a causal agent of stem rot of groundnut. *Karnataka Journal of Agricultural Sciences,* 7:365-367.

Kumar, B., Pandey, R. and Verma, A. K. 2003. Studies on host range of *Sclerotinia sclerotiorum* of broccoli. *Progressive Agriculture,* 3(1/2):131-132.

Kumar, N.G., Nirmala, P. and Jaayappa, A.H. 2010. Effect of various methods of application of insecticides on the incidence of serpentine leaf miner, *Liriomyza trifolii* (Burgess) and other pests in soybean. *Karnataka Journal of Agricultural Sciences*. 23(1):130-132.

Kumar, S. and Ray, P. 2011. Evaluation of augmentative release of *Zygogramma bicolorata* for biological control of Parthenium. *Crop Protection,* 30:587-591.

Kumar, S., Dhingra, A. and Daniell, H. 2004. Plastid-expressed *betainealdehyde dehydrogenase* gene in carrot cultured cells, roots, and leaves confer enhanced salt tolerance. *Plant Physiology,* 136(1):2843–2854.

Kushwaha, KS., Pareek, BL and Noor A. 1973. Fruit fly damage in cucurbits at Udaipur. *Udaipur University Research Journal*. 11:22–23.

Laghetti, G. and Hammer, K. 2007. The Corsican citron melon [*Citrullus lanatus* (Thunb.) Matsumura & Nakai subsp. lanatus var. citroides (Bailey) Mansf. Ex Greb.] A traditional and neglected crop. *Genetic Resources and Crop Evolution,* 54:913–916.

Lakshminarayana, M., Basappa, H. and Vijayasingh, R. 1992. Report on the incidence of hitherto unknown leaf miner (Agromyzidae) on castor. *J. Oilseeds Res.,* 9:175-176.

Lall, B.S. and Sinha. 1959. On the biology of the melon fly, *Dacus cucurbitae* Coq. (Diptera: Trypetidae). *Sci. & Cult.,* 25(2): 159-161.

Lall, H. and Singh, S. 1969. Studies on the biology and control of melon fly, *D. cucurbitae* (Diptera: Tephritidae). *Journal of Science and Technology,* 7B:148-153.

Lange, L., Eden, U. and Olson, L. W. 1989. Zoosporogenesis in *Pseudoperonospora cubensis.* The causal agent of cucurbit downy mildew. *Nordic Journal of Botany,* 8: 497-504.

Lati, R. N., Filin, S. and Eizenberg, H. 2011. Temperature- and radiation-based models for predicting spatial growth of purple Nutsedge (*Cyperus rotundus*). Weed Science, 59(4):476-482.

Laxminarayana, P. and Reddy, S. M. 1976. Post-harvest diseases of some cucurbitaceous vegetables from Andhra Pradesh. *Indian Phytopathology* 29: 57-59.

Lebesa, L. N., Khan, Z. R.; Hassanali, A.; Pickett, J. A.; Bruce, T. J. A.; Skellern, M. and Kruger, K. 2011. Responses of the blister beetle *Hycleus apicicornis* to visual stimuli. *Physiol. Entomol.,* 36:220-229.

Lee, K. Y., Lee, B. C. and Park, H. C. 1990. Occurrence of cucumber green mottle mosaic virus disease of watermelon in Korea. *Korean Journal of Plant Pathology,* 6:250-255.

Lee, S. and Yoo, J. G. 2006. Method for preparing transformed *Luffa cylindrica* Roem (World Intellectual property organization) http://www.wipo.int/pctdb/en/wo.jsp.

Lefroy, H. M. 1910. Life history of Indian Insects, Coleoptera-I. *Memoirs of Department of Agriculture in India, Entomological Series,* 2(8):139-164.

Leibee, G. L. 1984. Influence of temperature on development and fecundity of *Liriomyza trifolii* (Burgess) (Diptera: Agromyzidae) on celery. *Environmental Entomology* 13:497-501.

Lindegren, J. E., Wong, T. T. and Melnnis, D. O. 1990. Response of Mediterranean fruit fly (Diptera: Tephritidae) to the entomogenous nematode *Steinernema feltiae* in field tests in Hawaii. *Environ. Entomol.,* 19:383-386.

Lipton, W. J. 1977. Ultraviolet radiation as a factor in solar injury and vein tract browning of cantaloupes. *Journal of American Soceity for Horticultural Science* 102, 32-36.

Lipton, W. J. and Wang, C. Y. 1987. Chilling exposure and ethylene treatment change the level of ACC in Honey Dew melons. *Journal of American Society for Horticulrural Science,* 112:109-112.

Lolas, C. P. 1986. Control of broomrape (*Orobanche ramosa*) in tobacco (*Nicotiana tabacum*). *Weed Science* 34:427-430.

Ludlum, C. T., Felton, G. W. and Duffey, S. S. 1991. Plant defenses: Chlorogenic acid and polyphenol oxidase enhance toxicity of *Bacillus thuringiensis* sub sp. *karstaki* to *Heliothis zea. J. Chem. Ecol.* 17: 217-237.

Mabberley, D. J. 2008. Mabberley's Plant-Book: A portable dictionary of plants, their classification and uses, pp 1040. Cambridge University Press.

Mahato, A., Mondal, B. Dhakre, D. S. and Khatua, D. C. 2014. *In vitro* sensitivity of *Sclerotium rolsfii* towerds some fungicides and botanicals. *Scholars academic Journal of Biosciences,* 2(7):467-471.

Mahmood, K. and Mishkatulah 2007. Population dynamics of three species of Bactrocera (Diptera: Tephritidae: Dacinae) in BARI, Chakwal (Punjab). *Pakistan Journal of Zoology,* 39(2):123-126.

Mahmood, K., Akram, M., Quamar, S., Naqvi, A. and Alam, M. M. 1974. Studies on interaction between bottle gourd mosaic virus and powdery mildew fungus, *Sphaerotheca fuliginae. Indian Phytopath.* 27:827-829.

Makdoomi, S. M. A. and Ishaq, M. 1970. Chemical control of *Aulacophora foveicollis* (Lucas) on pumpkin crop. *Ann. Rep. Ayub Agric. Res. Inst.,* Faisalabad.

Mandal, N. C and Dasgupta, M. K. 1981. Post harvest diseases of perishables in West Bengal. Losses. *Ann. Agric. Res.,* 2:73-75.

Mandal, N. C. 1981. Post harvest disease of fruits and vegetable in West Bengal. Ph.D. Thesis, Visva - Bharathi, Santhiniketan.

Mandal, N. C. and Dasgupta, M. K. 1981. *Ann. Agric. Res.* 2:78-85.

Mandal, N. C. and Dasgupta, M. K. 1982. Post harvest disease of Perishables in West Bengal. India. *Indian Phytopath.,* 35:645-649.

Mandal, N. C. and Dasgupta, M. K. 1983. New report on post harvest diseases of perishable fruit and vegetables and their implication. *Geohios. New Rep.* 2:119-121.

Mandal, S., Mandal, B., Mohad, Q., Haq, R. and Varma, A. 2008. Properties, diagnosis and management of cucumber green mottle mosaic virus. *Plant Viruses,* 2(1):25-34.

Mandel, H., Levy, N., Izkovitch, S. and Korman, S. H. 2005. Elevated plasma citrulline and arginine due to consumption of watermelon (*Citrullus vulgaris*). *Journal of Inherited metabolic Disease,* 28(4):467–472.

Mani, M. S. 1934. "Studies on Indian Itonididae (Cecidomyiidae: Diptera)", *Rec. Indian Mus.,* 36:371-451.

Mani, M.S. 1973. Plant galls of India. McMillan India Ltd., Madras, pp-354.

Manjoo, S. and Swaminathan, R. 2007. Bio-ecology and management of *Henosepilachna vigintioctopunctata* (Fabricius) (Coleoptera: Coccinellidae) infesting ashwagandha [*Withania somnifera* (L.) Dunal]. *Journal of Medicinal and Aromatic Plant Sciences,* 29:16-19.

Masiunas, J. 2000. Weed control for commercial vegetable crops, In: Illinois Agricultural Pest Management Handbook. pp-197-225.

Matanmi, B.A. 1975. The biology of tephritid fruit flies (Diptera Tephritidae) attacking cucurbits at Ife Ife, Nigeria. *Niger J Entomol* 1:153–159.

Mathur, R. L. and Mathur, B. L. 1958. Fruit rot of *Citrullus vulgaris* var. *fistulosus* L. *Indian Phytopathology* 11:66-69.

May, A. W. S. 1946. Pests of Cucurbit Crops. *Queensland Agricultural Journal,* 62(3):137-150.

Mayee, C. D., Nandpuri, K. S. and Lal, T. 1976. A mosaic disease of muskmelon in Punjab. *Veg. Sci.,* 3:87-90.

Mc Phail, M. 1937. Relation of time, day temperature and evaporation to attractiveness of fermenting sugar solution to maxican fruit fly. *Journal of Economic Entomology,* 30:793-799.

McBride, O.C. and Tanda, Y. (1949). A revised list of host plants of the melon flies in Hawaii. Proceedings of the Hawaiian. Ento. Soc. 13: 411-421.Mahato, A. and Mondal, B. 2014. *Sclerotium rolsfii*: its isolates variability, pathogenicity and an eco-friendly management option. *Journal of Chemical, Biological and Physical Sciences* (Scetion B: Biological Sciences), 4(4): 3334-3344.

Meade, T. and Hare, J. D. 1993. Effects of differential host plant consumption by *Spodoptera exigua* (Lepidoptera: Noctuidae) on *Bacillus thuringiensis* efficacy. *Environ. Entomol.* 22: 432-437.

Mehrotra, B. S. 1954. The 'cottony leak' (*Pythium aphanidermatum*) of *Citrullus fistulosus* fruits. *Sci. Cult.* 19:564-565.

Mehrotra, R. S. 1980. Plant Pathology. Tata McGraw-Hill Publishing Company Limited, New Delhi, pp-771.

Minkenberg, O. P. J. M. 1988. Life history of the agromyzid fly *Liriomyza trifolii* on tomato at different temperatures. *Entomologica Experimentalis et Appliciata* 48:73-84.

Mishra, M. D., Lene, L. and Giri, B. K. 1983. Seed transmission of a white fly transmitted Gemini virus. *Proc. Fourth Int. Congr. Plant Path.* Melbourne (Abstr.).

Misra, S. and Gupta, N. 1988. 'Yellow' disease of vegetable and oil seed crops in Rajasthan. *Indian Phytopathology*, 41: 266 (abs.).

Mital, S. and Akram, M. 1985. *Cucumis sativus* (cucumber): a common host for *Erysiphe cichoracearum* DC and *Sphaerotheca fuliginea* (Schlecht.). *Poll. Acta Bot. Ind.* 13: 259-260.

Mondal, B. and Khatua, D. C. 2013. Evaluation of Plaster of Paris and some fungicides for management of foot rot of *Amorphophallus campanulatus* Blume caused by *Sclerotium rolfsii* Sacc. *International Journal of Agriculture, Environment & Biotechnology*, 6(4): 585-589.

Mondal, B., Bhattacharya, I. and Khatua, D. C. 2011. Crop and weed host of *Ralstonia solanacearum* in West Bengal. *Journal of Crop and Weed*, 7(2): 195-199.

Mondal, B., Bhattacharya, I. and Khatua, D. C. 2014. Incidence of bacterial wilt disease in West Bengal, India. *Academia Journal of Agricultural Research*, 2(6):139-146.

Mondal, B., Bhattacharya, R., Ranjan, R. K. and Khatua, D. C. 2004. Bacterial wilt disease of horticultural plants in West Bengal and its chemical control. *Green Technology*, 6:70-74.

Mondal, B., Das, R., Saha, G. and Khatua, D. C. 2014. Downy mildew of pointed gourd and its management. *Scholars Academic Journal of Biosciences*, 2(6): 389-392.

Mondal, B., Mahapatra, S. and Khatua, D. C. 2012. Records of some new diseases of horticultural plants of West Bengal. *Journal of Interacademicia,* 16(1): 36-43.

Mondal, B., Saha, G. and Khatua, D. C. 2013. Fruit and Vine Rot of Pointed Gourd in West Bengal. *Research Journal of Agricultural Sciences,* 4(1):44-47.

Motes, J. E. 1977. Pickling Cucumbers - Production and Harvesting. M.S.U. *Cooperative Ext. Service Bulletin,* E-837, pp-8.

Mukherjee, N. and Khatua, D. C. 1998. Diagnosis and management of vegetable bacterioses. In: Proceedings of National seminar on vegetable diseases and their management strategies in West Bengal (Sengupta, P. K. and Chowdhury, A. K. eds.). Department of Plant Pathology, Bidhan Chandra Krishi Viswavidyalaya. pp-48-63.

Mukherjee, S. K., Sharma, B. D. 1973. Root knot disease of *Trichosanthes dioica. Indian Phytopathology,* 26:248-249.

Mukhopadhyay S. and Saha K. 1968. Transmission of cucumis virus (cucumber mosaic virus) through seeds of *Cucurbita maxima* L. *Science and Culture,* 34 (N210):436-437.

Mukhopadhyay, S. and Nath, P. S. 1998. Current status of virus diseases of vegetables and research in West Bengal. In: Proceedings of National seminar on vegetable diseases and their management strategies in West Bengal (Sengupta, P. K. and Chowdhury, A. K. eds.). Department of Plant Pathology, Bidhan Chandra Krishi Viswavidyalaya. pp-64-72.

Mukhtar, I, Mushtaq, S., Khokhar, I. and Hannan, A. 2013. First record of *Cercospora citrullina* leaf spot on *Lagenaria siceraria* in Pakistan. *The Journal of Animal & Plant Sciences,* 23(6):1756-1757.

Mundkur, B. B. 1937. Anthracnose of cucurbits in Punjab. *Current Science,* 12:647-648.

Munro, M.K. 1984. A taxonomic treatise on the Dacidae (Tephritoidea, Diptera) of Africa. Republic of South Africa Department of Agriculture and water supply. *Entomological memoir* 61: 1-313.

Murata, M., Iwabuchi, K. and Mitsuhashi, J. 1994. Partial rearing of phytophagous lady beetle, *Epilachna vigintioctopunctata* (Coleoptera: Coccinellidae). *Zool Appl. Entomol.,* 29(1):116-119.

Mustafee, T. P. 1998. Fungicidal impact on vegetable disease management. In: Proceedings of National seminar on vegetable diseases and their management strategies in West Bengal (Sengupta, P. K. and Chowdhury, A. K. eds.). Department of Plant Pathology, Bidhan Chandra Krishi Viswavidyalaya. pp-77-82.

Nadagouda, S., Patil, B.V., Venkateshalu, and Sreenivas, A.G. 2010. Studies on development of resistance in serpentine leaf miner, *Liriomyza trifolii* (Burgess) (Agromyziidae; Diptera) to insecticides. *Karnataka Journal of Agricultural Sciences*. 23(1):56-58.

Nagaich, B. B. and Singh, B. 1960. Damping off of cucurbitaceous vegetables and its control. *Agra Univ. J. Res. (Sci.)*, 9: 125-135.

Nagendran, K., Aravintharaj, R., Mohankumar, S., Manoranjitham, S. K., Naidu, R. A. and Karthikeyan, G. 2015. First Report of *Cucumber green mottle mosaic virus* in Snake Gourd (*Trichosanthes cucumerina*) in India. *Plant Disease*, 99 (4):599.

Nagia, D. K., Kumar, S., Sharma, P., Meena, R. P., Saini, M. L. and Goel, S. C. 1992. Laboratory evaluation of insecticides for the control of *Henosepilachna vigintioctopunctata* (Fab.) on brinjal (*Solanum melongena* L.) (Coleoptera: Coccinellidae). *In*: Proceedings of the National Symposium on "Growth, Development and Control Technology of Insect Pests", Uttar Pradesh Zoological Society, Muzaffarnagar, India, pp-188-191.

Nair, M. R. G. K. 1995. Insects and mites of crops in India. Indian Council of Agricultural Research (ICAR), New Delhi, pp-165-166.

Nakahara, L.M. 1982. Notes and Exhibitions. *Proc. Hawaii. Entomol. Soc.* 24:8.

Narain, A., Swain, N. C., Sahoo, K. C., Das, S. K. and Shukla, V. D. 1985. A new leaf blight and fruit rot of watermelon. *Indian Phytopathology* 38:149-151.

Narayanan, E. S. 1953. The red pumpkin beetle and its control. *Indian Farming*, 3(2):8-9.

Narayanan, E. S. and Batra, H. N. 1960. *Fruit Flies and Their Control*. Indian Council of Agricultural Research, New Delhi, India, pp-1-68.

Nasr-Esfahana, M. and Ahmadi, A.R. 1997. *Applied Entomology and Phytopathology*, 65:18-20.

Nath, P. and Bhusan, S. 2006. Evaluation of poison bait traps for trapping adult fruit flies. *Annals of Plant Protection Sciences*, 14(2): 297-299.

Nath, P. S. 2004. Disease management in vegetable crops. In: Plant Pathology: Problems and Perspectives (Raj, S. K., Pan, S. K. and Chattopadhyay, S. B. eds.). pp-248.

Nath, R. P., Haidar, M. G., Akhter, S. W., Prasad, H. 1976. Studies on the nematodes of vegetables in Bihar-I. Effect of the reniform nematode, *Rotylenchulus reniformis* on *Trichosanthes dioica*. *Indian Journal of Nematology*, 6(2):175-177.

Nayar, N. N. and More, T. A. 1988. Cucurbits. Science Publishers, pp-340.

Nikbakhtzadeh, M. R. 2004. Transfer and distribution of cantharidin within selected members of blister beetles (Coleoptera: Meloidae) and its probable importance in sexual behaviour. Ph.D. dissertation. Universitat Bayreuth, Germany, pp-105.

Nishijima WT, Aragaki M 1993. Pathogenicity and further characterization of *Calonectria crotalariae* causing collar rot of papaya. *Phytopathology,* 63: 553–558.

Noble, D. 2009. Working toward better cucurbits: Weed management strategies for the coming season (http://www.growingmagazine.com/fruits/working-toward-better-cucurbits/(http://www.growing magazine.com/fruits/working-toward-better-cucurbits/) .

Oerke, E. C., Dehne, H. W., Schonbeck, F. and Weber, A. 1994. Crop Production and Crop Protection: Estimated Losses in Major Food and Cash Crops, Elsevier Publishing Co., Amesterdam, The Netherlands. pp-1-44.

Ojha, N. L. 1983. Ph. D. thesis submitted to Bhagalpur University, Bhagalpur-812007, India.

Okada, Y. 1986. Cucumber green mottle mosaic virus. *In*: The Plant Viruses (Van Regenmortel *et al* . eds.), Plenum Press, New York, pp-267-281.

Okinawa Prefecture, 1987. Melon fly eradication project in the Okinawa Prefecture. Okinawa Prefecture fruit fly eradication Project Office, Naha.

Okonmah, L.U. 2011. Effect of different types of staking and their cost effectiveness on the growth, yield and yield components of cucumber (*Cumumis sativa* L). *International Journal of Agriculture Science,* 1(5):290-295.

Olliff, A. S. 1980. The leaf-eating lady-bird, *Epilachna vigintioctopunctata* Fabr. *Agric. Gaz. N.S.W.* 1:281-283.

Ordonez, A. A., Gomez, J. D. and Isla, M. A. 2006. Antioxidant activities of *Sechium edule* (Jacq.) Swartz extracts. *Food Chemistry,* 97:452-58.

Orian, A. J. E. and Moutia, L. A. 1960. Fruit flies (Trypetidae) of economic importance in Mouritius. *Revue Agricole et Sucriere de Ille Mourice,* 39: 142-150.

Packauskas, R. J. and Schaefer, C. W. 2001. Clarifications of some taxonomic problems in Anisoscelini and Leptoscelini (Hemiptera: Coreidae: Coreinae). *Proc. Entomol. Soc. Washington* 103:249-256.

Pagden, H. T. 1928. Leptoglossus membranaceus F., A pest of cucurbitacae. *Malaya Agricultural Journal,* 16(2):387-403.

Palodhi, P. R. and Sen, B. 1979. Role of tylose development in the muskmelon disease caused by *Fusarium solani. Pl. Dis. Reptr.* 63:584-586.

Panda, N. 2003. Host plant resistance and insect pest management. Proceedings of the "National Symposium on Frontier Areas of Entomological Research", November, 5-7, 2003, IARI, New Delhi, India, pp-377-394.

Pandey, M.B. and Misra, D.S. 1999. Studies on movement of *Dacus cucurbitae* maggot for pupation. *Shashpa* 6:137-144.

Pandey, K. K., Pandey, P. K., Satpathy, S. 2002. Integrated management of diseases and insects of tomato, chilli and cole crops. *Indian Institute of Vegetable Research, Varanasi, Tech. Bull.* No. 9:1-22.

Panji, H. R. 1964. Some observations on the insecticidal activities of the fruit of 'Darek', *Melia azedarach* (L.). *Research Bulletin of Punjab University*, 15:4345-4346.

Paolo, M. A. 1933. A sclerotium seed rot and seedling stem rot of mango. *Philippine Journal of Science*, 52:237-261.

Pareek, B.L. and V.S. Kavadia, 1988. Economic Entomology, 29: 349-352. Insecticidal control of two major pests of musk melon, 33. Lewis, P.A. and R.L. Metcalf, 1996. Behavior and *Cucumis melo* in the pumpkin beetle, Raphidopalpa ecology of Old World Luperini beetles of the genus spp. and the fruitfly, Dacus cucurbitae in Rajasthan, Aulacophora (Coleoptera: Chrysomelidae). *Journal of Tropical Pest Management*. 34(1):15-18.

Parle, M. and Singh, K. 2011. Musk melon is eat-must melon: A Review. *International Research Journal of Pharmacy*, 2(8):52-57.

Parrella MP. 1987. Biology of *Liriomyza*. *Annual Review of Entomology*, 32: 201-224.

Parrella, M. P., Yost, J. T.; Heinz, K. M. and Ferrentino, G. W. 1989. Mass rearing of *Diglyphus begini* (Hymenoptera: Eulophidae) for biological control of *Liriomyza trifolii* (Diptera: Agromyzidae). *Journal of Economic Entomology* 82: 420-425.

Patel, K. N. and Purohit, M. S. 2000. Host preference of epilachna beetle, *Epilachna vigintioctopunctata* Fab. *Gujrat Agricultural University Research Journal*, 25(2): 94-95.

Patnaik, H. P., Sarangi, P. K. and Mahapatra, P. 2004. Studies on the incidence of fruit flies and jassids on summer bitter gourd and their control. *Orissa Journal of Horticulture*, 32(2):87-90.

Peck, S. L. And McQuate, G. T. 2000. Field tests of environment friendly malathion to suppress wild Mediterranean fruit fly (Diptera: Tephritidae) populations. *Journal of Economic Entomology*, 93:280-290.

Pedigo, L. P. 1995. Entomology and Pest Management. Macmillan Publishing Company, New York. pp-107-119.

Pedigo, L. P. 2002. Entomology and pest management. Prentice Hall of India Pvt. Ltd., New Delhi, India, pp-441-476.

Philip, A. 1950. Description of one new species of *Strumeta* Walker. (Trypetidae: Diptera) from Burma and a record of one far Eastern species of the genus from India. *Indian Journal of Entomology*, 10:31-32.

Picha, D. 1986. Postharvest fruit conditioning reduces chilling injury in watermelons. *Hort. Science* 21:1407-1409.

Pillai, M. G. 1971. A mosaic disease of snake gourd (*Trichosanthes anguinal*). *Sci. & Cult.* 37:46-47.

Pitcha, D. H. 1986. Post harvest fruit conditioning reduces chilling injury in watermelons. *Hort Science,* 21:1407-1409.

Pitchaimuthu, M., Souravi, K., Ganeshan, G., Kumar, G. S. and Pushpalatha, R. 2012. Identification of sources of resistance to powdery and downy mildew diseases in Cucumber [*Cucumis sativus* (L.)]. *Pest Management in Horticultural Ecosystems,* 18 (1):105-107.

Plumb, R. T., Mukhopadhyay, S. and Jones, P. 2000. Viruses of crops, weeds in Eastern India. IACR-Rothamsted, Hertfordshire United Kingdom and Bidhan Chandra Krishi Viswavidyalaya, India.

Pluthero, F. G. and Singh, R. S. 1984. Insect behavioural response to toxins: Practical and evolutionary consideration. *Canadian Entomologist,* 116: 57-68.

Polischuk, V., Budzanivska, I., Shevchenko, T. and Oliynik, S. 2007. Evidence for plant viruses in the region of Argentina Island, Antarctica. *FEMS Microbilolgy and Ecology,* 59: 409-417.

Pradhan, S., Jotwani, M. G. and Prakash, S. 1990. Comparative toxicity of insecticides to the grub and adult of *Epilachna vigintioctopunctata* Fab. (Coleoptera: Coccinellidae). *Indian J. Ent.* 24(4):223.

Prakash, O. and Singh, S. J. 1977. Anthracnose of bitter gourd caused by *Colletotrichum lagenarium. Indian Phytopath.,* 30:290-291.

Prasad, M. M. and Poddar, K. D. 1977. Physiological changes in vegetables during pathogenesis. In: Bilgrami, K. S. (ed.) Physiology of microorganism. Today and Tomorrow's Printers Publishers, New Delhi, pp-275-284.

Prasad, S. S. and Ambasta, K. K. 1987. Fruit rot of sponge gourd. *Indian Journal of Mycology and Plant Pathology,* 17:235.

Price, J. F. and Poe, S. L. 1976. Response of *Liriomyza* (Diptera: Agromyzidae) and its parasites to stake and mulch culture of tomatoes. *Florida Entomologist,* 59:85-87.

Prokopy, R. J., Miller, N. W., Pinero, J. C., Barry, J. D., Tran, L. C., Oride, L. K. and Vargas, R. I. 2000. Effectiveness of GF-120 Fruit Fly Bait spray applied to border area plants for control of melon flies (Diptera: Tephritidae). *Journal of Economic Entomology*, 96:1485-1493.

Punja, Z. K. 1985. The biology, ecology and control of *Sclerotium rolfsii. Annual Review Phytopathology,* 23:97-127.

Punja, Z. K. 1988. Genetic of plant pathogenic fungi (Sidhu, G. S. ed.). Academic Press, London, 6:523-534.

Purchifull, D., Edwardson, J., Hiebert, E. and Gonsalves, D. 1984. Papaya ringspot virus. CMI/AAB Description of Plant Viruses No. 292. CMI, Kew, U. K.

Qureshi, Z. A., Bughio, A. R.; Siddiqui, Q. H. and Najibullah 1976. Efficacy of methyl eugenol as a male attractant for *Dacus zonatus* (Saunders) (Diptera: Tephritidae). *Pakistan Journal of Scientific and Industrial Research,* 19:22-23.

Rabindranath, K and Pillai, K.S. 1986. Control of fruit fly of bitter gourd using synthetic pyrethroids. *Entomon*. 1986;11:269–272.

Radhakrishnan, P. and Sen, B. 1985. Effect of different methods of inoculation of *Fusarium oxysporum* and *Fusarium solani* for inducing wilt of muskmelon. *Indian Phytopathology,* 38:70-73.

Rai, A.B., Loganathan, M., Halder, J., Venkataravanappa, V. and Naik, P. S. 2014. Eco-friendly Approaches for Sustainable Management of Vegetable Pests. IIVR Technical Bulletin-53, *Published by Indian Institute of Vegetable Research,* Varanasi, Uttar Pradesh-221305. pp-104.

Rai, M., Pandey, S. and Kumar, S. 2005. Cucurbit research in India: a retrospect. *Indian Institute of Vegetable Research,* Jakhini (Shahanshahpur) Varanasi, India, pp-285-294.

Rai, M., Pandey, S. and Kumar, S. 2008a. Cucurbit research in India: a retrospect. Cucurbitaceae 2008, Proceedings of the IX EUCARPIA meeting on genetics and breedingof Cucurbitaceae (Pitrat, M. ed), INRA, Avignon (France), May 21-24th, 2008.

Rai, M., Pandey, S. and Kumar, S. 2008b. Cucurbitaceae. Proceedings of the IXth EUCARPIA meeting on genetics and breeding of Cucurbitaceae (Pitrat, M., ed), INRA, Avignon (France), May 21-24th, 2008.

Rai, N., Sanwal, S. K., Yadav, R. K. and Phukan, R. M. 2006. Diversity in Chow-chow in north eastern region. *Indian Horticulture,* 51(2):11-12.

Rai, N., Yadav, D. S., Nath, A. and Yadav, R. K. 2002. Chowchow: a poor man vegetable for north eastern hills region. *Indian Farming,* 34:18-20.

Rajagopal, D. and Trivedi, T. P. 1989. Status, bioecology and management of epilachna beetle, *Epilachna vigintioctopunctata* (Fab.) (Coleoptera :

Coccinellidae) on potato in India: A review. *Tropical Pest Management*, 35(4):410-413.

Rajendran, V. 1965. A note on the occurrence of perfect stage of bottle gourd powdery mildew in Mysore. *Indian Phytopathology,* 18:389-390.

Ram, D., Kalloo, G. and Banerjee, M. K. 2002. Popularizing kakrol and kartoli: the indigenous nutritious vegetables. *Indian Horticulture,* 47(3):6-9.

Ram, D., Rai, M., Rai, N., Yadav, D. S., Pandey, S., Verma, A., Lal, H., Singh, N. and Singh, S. 2006. Characterization and evaluation of winter fruited bottle gourd, (*Lagenaria siceraria* Mol.) *Acta Horticulturae,* 752:231-238.

Ramakrishna, Ayyar, T. V. 1920. "Some Insects recently noted as Injurious in South India", *Proc. 3rd Ent. Mtg. Pusa,* 1:314-328.

Ramandeep, K. and Mavi, G. S. 2005. Biology of *Epilachna vigintioctopunctata* (Fabricious) (Coleoptera: Coccinellidae) on brinjal in Ludhiana, Punjab. *Crop Research,* 29(1):141-144.

Ramaswamy, K. and Prasad, N. N. 1975. Effect of potassium nutrition on phenol metabolism of melon wilt. *Madras Agri. J.,* 62:313-317.

Ramsey, G. B. and Smith, M. A. 1961. Market diseases of cabbage, cauliflower, turnips, cucumbers, melons and related crops. *U. S. Dept. Agric. Handbook.* 184:49.

Ramzan, M., S. Darshan, G. Singh, G.S. Mann and J.S. Bhalla, 1990. Comparative development and seasonal abundance of Hadda beetle, *H. vigintiocto-punctata* (Fabr.) on some solanaceous host plants. *J. Res. Punjab Agric. Univ.,* 27: 253-262.

Randall, P. 2012. A Global Compendium of Weeds. 2nd Edition. Department of Agriculture and Food, Western Australia.

Ranganath, H. R., Suryanarayana, M. A.; Veenakumari, K. 1997. Management of melon fly, (*Bactrocera* (*Zeugodacus*) *cucurbitae* on cucurbits in South Andaman. *Insect Environment,* 3:32-33.

Rangaswami, G. and Mahadevan, A. 2004. Diseases of crop plants in India. 4th Ed., Prentice-Hall of India Pvt. Ltd., New Delhi, pp-536.

Rani, S., Sharma, G. S. and Verma, H. N. 1971. Epidemiology of virus diseases of cucurbits. *Proc. Indian Nat. Sci. Acad.* 37:345-351.

Rao, A. L. N. and Varma, A. 1984. Transmission studies with cucumber green mottle mosaic virus. *Phytopathologische Zeitschrift,* 109:325-331.

Rao, D. R. 1976. Epidemiology and characterization of viruses naturally occurring on bottle gourd and vegetable marrow. Ph. D. Thesis, PG School, IARI, New Delhi.

Rao, M. H. P., Raychaudhuri, S. P. and Varma, A. 1976. Inhibition of cucumber mosaic virus by some chemicals. *Act Phytopath. Acad. Sci. Hung.* 11: 259-269.

Rao, V. G. 1964. Some new market and storage disease of fruits and vegetables in Bombay (Maharashtra). *Mycopath. Mycol. Appl.* 23:297-310.

Rao, V. G. 1965. Some new records of market and storage diseases of vegetable in bombay (Maharashtra). *Mycopath. Mycol. Appl.* 27: 49-59.

Rao, V. G. 1966. An account of the market and storage diseases of fruits and vegetables in Bombay (Maharashtra). *Mycopath. Mycol. Appl.,* 28: 165-176.

Rashid, M. A., Khan, M. A., Arif, M. J and Javed, N, 2015. Intensive Management of Red Pumpkin Beetle (*Aulacophora foveicollis* Lucas) in Different Ecological Regions. *Pakistan J. Zool.,* 47(6): 1611-1616.

Raychaudhuri, M. 1973. Studies on cucurbit viruses. Ph.D. Thesis. PG School, IARI, New Delhi.

Raychaudhuri, M. and Varma, A. 1975a. Virus diseases of cucurbits in Delhi. Proc. 62nd Indian Sci. Congr. Part III. pp-74.

Raychaudhuri, M. and Varma, A. 1975b. Virus vector relationship of marrow mosaic virus with *Myzus persicae* Sulz. *Indian J. Ent.* 37:243-250.

Raychaudhuri, M. and Varma, A. 1978. Mosaic disease of muskmelon, caused by a minor variant of cucumber green mottle mosaic virus. *Phytopathologische Zeitschrift,* 93:120-125.

Reddy, H. R. and Nariani, T. K. 1963. Studies on mosaic disease of vegetable marrow (*Cucurbita pepo* L.). *Indian Phytopath.* 16:260-267.

Renjhan PL. 1949. On the morphology of the immature stages of *Dacus* (*Strumeta*) *cucurbitae* Coq. (the melon fruit fly) with notes on its biology. *Indian Journal of Entomology.* 11:83–100.

Richards, 1983. The *Epilachna vigintioctopunctata* complex (Coleoptera: Coccinellidae). *International Journal of Entomology,* 25(1):11-41.

Risse, L. A. and Hatton, T. J. 1982. Sensitivity of water melon to ethylene during storage. *Hort Science,*17:946-948.

Robak, J. 1995. Epidemiology and control of cucumber downy mildew *Pseudoperonospora cubensis. Biuletyn Warzywniczy* 43:5-18.

Robinson, R. W. and Decker-Walters, D. S. 2004. Cucurbits: Crop Production Science in Horticulture, 6. CAB Publishing, Wallingford, UK.

Rohani, I. 1987. Identification of larvae of common fruit fly pest species in West Malaysia. *Journal of Plant Protection in the Tropics,* 4:135-137.

Roy, A. K. and Singh, H. N. P. 2003. Appraisal of post-harvest pathology of perishables (fruits and vegetables) in India. In: Plant Pathology at Crossroads (Sen, C., Dasgupta, M. K. and Chaudhuri, S. Eds.)., B. C. K. V., Mohanpur, pp-270-283.

Roy, A. K. 1973. Fruit rot of bottle gourd. *FAO Plant Protection Bulletin,* 21(5): 116-117.

Roy, S. and Mukhopadhyay, S. 1979. Epidemiology of virus disease of pumpkin in West Bengal. *Indian Phytopathology,* 31(2):207-209.

Roychaudhury, S. P. and Verma, J. P. 2000. Diseases of crops. In: Handbook of Agriculture. Directorate of Information and Publications of Agriculture. ICAR, Krishi Anusandhan Bhavan, Pusa, New Delhi, pp-1303.

Rukmani, R., Nidhiya, I. S. R., Nair, S. and Kumar, A. 2003. Investigation of anxiolytic-like effect of antidepressant activity of *Benincasa hispida,* methanol extract. *Indian Journal of Pharmacology,* 35:129-130.

Ryall, A. L. and Lipton, W. J. 1972. Handling, transportation and storage of fruits and vegetables. Vol. 1, AVI Pub. Co., Westport CT.

Ryu, K. H., Min, B. E., Chol, G. S., Chol, S. H., Kown, S. B., Noh, G. M., Yoon, J. Y., Chol, Y. M., Jang, S. H., Lee, G. P., Cho, K. H. and Park, W. M. 2000. zucchini green mottle mosaic virus is a new tobamovirus: comparison of its coat protein gene with that of Kyuri green mottle mosaic virus. *Archives of Virology,* 145:2325-2333.

Saha, G. 2002. Fruit and vine rot of pointed gourd. Ph. D. thesis, Bidhan Chandra Krishi Viswavidyalaya, Mohanpur, West Bengal, pp. 147.

Saha, G., Das, S. and Khatua, D. C. 2004. Fruit and vine rot of pointed gourd. *Journal of Mycopathological Research,* 42:73-81.

Saha, G., Maity, S. S. and Khatua, D. C. 2002. *Journal of Mycopathological Research,* 40(2): 145-147.

Sahu Kritagyan, S. P. and Singh, S. P. 1980. Fruit rot of pointed gourd (*Trichosanthus dioica*) in Bihar. *Ind. Phytopathol.* , 33:308-309.

Saleem, M.A. and H.A. Shah, 2010. Applied Entomology 3 edition Publisher rd Pak Book Empire 712 Sector A-I Town Ship Lahore, pp-266-267.

Saljoqi, A. U. R. and Khan, S. 2007. Relative abundance of the red pumpkin beetle, *Aulacophora foveicollis* (Lucas) on different cucurbitaceous vegetables. *Sarhad Journal of Agriculture,* 23(1):109-114.

Samalo, A. P. 1985. Chemical Control of Pointed Gourd Vine Borer, (*Apomecyna saltator*) Fabr. *Madras Agric. J.* 72 (6): 325-329.

Samalo, A. P. and Parida, P. B. 1983. Influence of Spacing and Levels of Nitrogen on Vine-Borer Incidence and Yield of Pointed-Gourd. *Indian J. Agric. Sci.* 53 (7):574-577.

Sanford, D., Eigenbrode and Trumble, J. T. 1994. Host Plant Resistance to Insects in Integrated Pest Management in Vegetable Crops. *Journal of Agricultural Entomology*, 11(3): 201-224.

Sarkar, B. B. and Mukhopadhyay, S. 1979. Separation of nine anisometric cucurbit viruses by host reactions. *Vegetable Science* 6(2):142-144.

Sarma, B. K. and Singh, U. P. 2002. *Mycologia*, 94(6):1051-1058.

Satoh, I., Yamabe, M., Satoh, S. and Ohki, A. 1985. Study of the frequency of finding of the fruit fly infesting the fruits imported as air bagging. *Research Bulletin of the Plant Protection Service*, Japan, 21:71-73.

Satpathy, S., Samarjit, R. and Kapoor, K. S. 1998. Integrated Management of Vegetable Pests. In: G. Kalloo (Eds.), National Symposium on "Emerging Scenarion in Vegetable Research and Development", 12-14 December, 1998. Indian Society of Vegetable Science, Project Directorate of Vegetable Research, Varanasi, India, pp-123-130.

Satyagopal, K., Sushil, S. N. and Jeyakumar, P. 2014. AESA based IPM – Cucurbitaceous Vegetable Crops (Cucumber, Bottle Gourd, Bitter Gourd, Sponge Gourd, Snake Gourd, Ash Gourd, Pumpkin, Squash). AESA BASED IPM Package No. 21, National Institute of Plant Health Management, Hyderabad, India.

Schuster, D. J.; Gilreath, J. P.; Wharton, R. A. and Seymour, P. R. 1991. Agromyzidae (Diptera) leafminers and their parasitoids in weeds associated with tomato in Florida. *Environmental Entomology*, 20:720-723.

Seebold, K. 2010. Foliar diseases of cucurbits. UK Cooperative Extension Service, University of Kuntucky, College of Agriculture. *Plant Pathology Fact Sheet* (PPFS-VG-10).

Selander, R. B. 1986. Rearing blister beetles (Coleoptera, Meloidae). *Insecta Mundi*. 1:209-220.

Seymour, P. R. 1994. Taxonomy and morphological identification. *In* EU Contract No. 90/399005 Final Report - Evaluation and development of rapid detection and identification procedures for *Liriomyza* species: taxonomic differentiation of polyphagous *Liriomyza* species of economic importance (Aukema B., Oudman L., Menken S. B. J., Ulenberg S. A., Seymour P. R. and Martinez M.).

Shaheen, A. H., Samhan, M. and Elezz, A. A. 1973. Cucurbit pests at Komombo. *Agriculture Research Review*, 51:97-101.

Shang, H., Xie, Y., Zhou, X., Qian, Y. and Wu, J. 2011. Monoclonal antibody-based serological methods for detection of Cucumber green mottle mosaic virus. *Virol. J.* 8:228.

Shankar, G. and Nariani, T. K. 1974. A mosaic disease of watermelon (*Citrullus vulgaris* Schrad.). *Curr. Sci.* 43:281-282.

Shankar, G., Nariani, T. K. and Prakash, N. 1971. Purification and serology of bottle gourd (*Lagenaria siceraria* Standl.) mosaic virus. *Indian J. Microbiol.,* 11: 43-52.

Shankar, G., Nariani, T. K. and Prakash, N. 1972. Studies on pumpkin mosaic virus (PMV) with particular reference to purification, electron microscopy and serology. *Indian J. Microbiol.,* 12: 154-165.

Sharaf, N. S. 1989. Monitoring spider mite populations on lemon for effective control with flubenzimine and omethoate. *Dirasat* 16:53-64.

Sharma, C., Trivedi, P. C. and Tiagi, B. 1985. *Internationa Nematology Network Newsletter,* University of Rajasthan, India, 2:7-9.

Sharma, O. P., Khatri, H. L., Bansal, R. D. and Komal, H. S. 1984. A new strain of cucumber mosaic virus causing mosaic disease of muskmelon. *Phytopath. Z.,* 109:332-340.

Sharma, P. D. 2005. Fungi and allied organisms. Narosa Publishing House Pvt. Ltd., New Delhi, pp-545.

Sharma, R. and Gupta, R. 2007. Cyperus rotundus extract inhibits acetylcholinesterase activity from animal and plants as well as inhibits germination and seedling growth in Wheat and Tomato. *Life Science,* 80: 24-25.

Sharma, Y. R. and Chohan, J. S. 1973. Transmission of Cucumis viruses-1 and 3 through seeds of cucurbits. *Indian Phytopathology,* 26:596-598.

Shinde, C. B. and Purohit, M. L. 1978. Population study of red pumpkin beetle, A. foveicollis, a serious pest of cucurbits. *Science and Culture,* 44(7):335.

Shukla, G. S. and Upadhyay, V. B. 1985. Management of *Epilachna dodecastigma* a vegetable pest. 1. Effect of humidity on the attractiveness. *Zeits-for Agewandte Zool.* 72(3):305-308.

Siddaramaiah, A. L., Kulkarni, S. and Hegde, R. K. 1982. A new fruit rot disease of pumpkin from India. *Indian Phytopathology,* 35:705-707.

Sikora, E. J. 2011. Common diseases of cucurbits. Publication of the Alabama Cooperative Extension System (Alabama A&M University and Auburn University), ANR-0809, pp. 1-8.

Sindhu, P. V., Thomas, C. G. and Abraham, C. T. 2010. Seedbed manipulations for weed management in wet- seeded rice. *Indian Journal of Weed Science,* 42(3&4):173-179.

Singh S.J. 1992. Electron microscopy observations of bitter gourd witches broom. *Indian Phytopath.,* 45:354-355.

Singh, S.V., Mishra, A., Bisan, R.S., Malik, Y.P and Mishra, A. 2000. Host preference of red pumpkin beetle, *Aulacophora foveicollis* and melon fruit fly, *Dacus cucurbitae. Indian Journal of Entomology.* 62:242–246.

Singh, B. K., Ramakrishna, Y. and Verma, V. K. 2015. Chow-Chow (*Sechium edule*): Best Alternative to Shifting Cultivation in Mizoram. *Indian Journal of Hill Farming*, 28(2):158-161.

Singh, D. V. and Seth, M. L. 1974. Collar rot of pumpkin caused by *Sclerotium rolfsii*. *Indian Phytopath*. 27:631-632.

Singh, D., Nandpuri, K. S. and Sharma, B. R. 1976. Inheritance of some economic quantitative characters in an intervarietal cross of muskmelon (*Cucumis melo* L.). *Res. Ludhiana*, 13:172-176.

Singh, H. S. and Naik, G. 2006. Seasonal dynamics and management of pumpkin caterpillar, *Diaphania indica* saunders and fruit fly, *Bactrocera cucurbitae* Conq. in bitter gourd. *Vegetable-Science*. 33(2): 203-205.

Singh, J. P. 1970. Elements of vegetable pests. Vora and Co. Publishers Pvt. Ltd. Bombay. Pp-275.

Singh, J. P. and and Gupta, R. 1970. The red pumpkin beetle, Aulacophora foveicollis (Lucas), as a pest of Japanese mint. *Journal of Bombay Natural History Soceity*, 67(1):123-124.

Singh, R. 2014. Weed management in major kharif and rabi crops. *Proceedings of National Training on Advances in Weed Management*. pp. 31-40.

Singh, R. S. 1986. A new fruit rot disease of muskmelon from India. *PAU Journal of Research*, 23: 265-266.

Singh, R. S. 1990. Plant Diseases. Oxford & IBH Publishing Company, New Delhi, pp-619.

Singh, R. S. and Chohan, J. S. 1972. Charcoal rot of ridge gourd - a pycnidial strain of *Macrophomina phuseoli* (Maubl.) Ashby from India. *Indian Phytopathology*, 25: 463-464.

Singh, R. S. and Chohan, J. S. 1977. Factors affecting growth and fruit rot (cottony leak) of cucurbits caused by *Pythium butleri*. *Indian Phytopathology*, 30:379-383.

Singh, R. S. and Chouhan, J. S. 1980. Fungal fruit rot of bottle gourd in North India. *Indian Phytopathology*, 33: 598-599.

Singh, R., Singh, R. K., Rai, R. K., Sharma, R. and Tiwari, S. P. 2013. Improvement of *Trichoderma* spp. strain for integrated control of soil borne plant pathogens. *International Journal of Agriculture, Environment & Biotechnology*, 6(3): 487-496.

Singh, S. J. 1973. A stem end rot disease of snake gourd fruits caused by *Glornerella cingutata* (Stonem.) Spauld. & Shrenk. *Current Science*, 42:296.

Singh, S. J. 1974. A fruit rot of snake gourd caused by *Alternuria tennuissirna*. *Indian Phytopathology*, 27:384-385.

Singh, S. J. 1981. Studies on virus causing mosaic disease of pumpkin (*Cucurbita maxima Duch.*). *Phytopath. Medit.* 20: 104.

Singh, S., Kumar, A. and Pandey, N. D. 2005. Relative toxicity of some commonly used insecticides against hadda beetle in brinjal. *Indian Journal of Entomology,* 67(1): 21-23.

Sinha, A. K. 2001. Fundamentals of Plant Pathology. Kalyani Publishers, New Delhi, pp. 183.

Sinha, P. P. 1990. Significance of chemical control of parwal powdery mildew. *Indian Phytopathology,* 43 (2): 229-230.

Slosser, J. E., W. E. Pinchak, and D. R. Rummel 1989. A review of known and potential factors affecting the population dynamics of the cotton aphid. *Southwest. Entomol.* 14: 302-313.

Smith-Meyer, M. K. P. 1981. Mite pests of crops in southern Africa. *Science Bulletin, Department of Agriculture and Fisheries, Republic of South Africa* No. 397: 65-67.

Sohi, H. S. 1975. Recent advances in the control of fungal diseases of horticultural crop plants in India. *Pesticides Ann. No.* 101-105.

Sohi, H. S. 1984. Present status of our knowledge of important fungal diseases of selected vegetables in India and future needs. Presidential Address. *Indian J. Mycol. Pl. Path.,* 14:1-34.

Sohi, H. S. and Nayar, S. K. 1969. Some new records of fungi from India. *Indian Phytopathology,* 22:410-421.

Sohi, H. S. and Sharma, S. R. 1998. Fungal diseases and their management. In: Cucurbits (Nayar, N. M. and More, T. A. eds.). Oxford and IBH Publishing Co. Pvt. Ltd., New Delhi, pp-211-245.

Sohi, H. S., Kaur, H. and Singh, G. 1981. Occurrence of cleistothecial stage of *Erysiphe cichoracearum* DC on *Coccinia indica* W. & A. *Current Science,* 50: 779-780.

Sohi, H. S., Ullasa, B. A. and Sokhi, S. S. 1976. Blossom end rot of bottle gourd (*Lagenaria siceraria* (Mol.) Standl.) incited by *Pythium butleri. Current Science,* 45:630-631.

Som, D. and Bandyopadhyay, A. K. 1981. *Rhizoctonia bataticola* (Taub.) Butlet on *Trichosanthes dioica* Roxb- A new report. *Current Science,* 50:192-193.

Sood, N. and Sharma, D. C. 2004. Bioefficacy and persistent toxicity of different insecticides and neem derivatives against cucurbit fruit fly, *Bactrocera cucurbitae* (Coq.) on summer squash. *Pesticide Research Journal,* 16(2):22-25.

Sounder Rajan, K. Dhandapani, N. and Chezhiyan, N. 1996. A low cost technology to control fruit fly, *D. cucurbitae* of chow chow. *Pestology,* 90: 15-16.

Spaugy, L. 1988. Fruit flies. Two more eradication projects over. *Citrograph* 73: 168.

Sreedevi, K., Prasad, K. V. H. and Srinivasan, S. 2009. Occurrence of orange banded blister beetle, *Mylabris pustulata* Thun. on Cashew apple in Tirupati region of Andhra Pradesh. *Current Biotica,* 3(3):450-451.

Srinivasan, K. and Pal, A. B. 1998. Pests and their management. *In* Cucurbits (Edt. By Nayar, N. M. and Pal, A. B.). Oxford and IBH Publishing Co. Pvt. Ltd. pp- 247.

Srinivasan, K., Viraktamath, C. A., Gupta, M. and Tiwari, G. C. 1995. Geographical distribution, host range and parasitoids of serpentine leaf miner, *Liriomyza trifolii* (Burgess) in south India. *Pest Mgmt. in Hort. Ecosys.,* 1:93-100.

Srivastava, K. P. 1996. A textbook of Applied Entomology, Vol-II, Kalyani Publishers, Ludhiana, India, pp-131.

Srivastava, K. P. and Butani, D. K. 2009. Pest Management in Vegetables. Vol.-II. Studium Press (India), New Delhi.

Srivastava, S. L. and Suman, B. 1986. Inhibitory effect of some phylloplane fungi on powdery mildew disease development. *Indian Phytopathology,* 39: 83-86.

Stegmaier, C. E. 1966. Host plants and parasites of *Liriomyza munda* in Florida (Diptera: Agromyzidae). *Florida Entomologist,* 49:81-86.

Stein, U., Parrella, M.P., 1985. seed extract shows promise in leafminer control. calif. agric.39 (7/8) 19–20 (c.f. r.a.e., a, 1993).

Stovel, D. D. 2005. Pumpkin: A Super Food for All 12 Months of the Year. North Adams, MA: Storey Publishing, LLC.

Subratty, A. H, Gurib-Fakim, A. and Mabomoodally, E. 2005. Bitter melon: An exotic vegetable with medicinal values. *Nutrition and Food science,* 35: 143-147.

Suhag, L. S. and Duhan, J. C. 1980. Collar rot of muskmelon, a new record. *Indian Phytopathology,* 33:94.

Suhag, L. S. and Mehta, N. 1982. Economics, efficacy and assimilation of some antipowdery mildew fungicides in cucurbit crops. *Indian Phytopathology,* 35:104-105.

Sumbali, G., and Mehrotra, R. S. 1982. Cucurbitaceous fruits –New hosts for Fusarium equiseti. *National Academy of Science Letters – India,* 5:121–122.

Suryanarayana, D., Ramnath and Lal, S. P. 1963. Seed borne infection of stackburn disease of rice its extent and control. *Indian Phytopathology,* 16: 232-233.

Swamy, K. R. M., Datta, O. P. and Ullasa, B. A. 1981. Identifying the sources of resistance to downy and powdery mildew diseases in muskmelon. In: *Proc.*

Nat. Sem. on Dis. Res. in Crop Plants, December 22-23. TNAU, Coimbatore, pp-101-104.

Swarup, V. 2006. Vegetable science and technology in India, Kalyani Publishers, New Delhi. pp- 426-431.

Syed, R. A. 1970. Studies on Trypetids and natural enemies in West Pakistan. Dacus species of lesser importance. *Pakistan Journal of Zoology*, 2:17-24.

Talpur, M. A., Rustamani, M. A.; Hussain, T; Khan, M. M. and Katpur, P. B. 1994. Relative toxicity of different concentrations of Dipterex and Anthio against melon fly, *Dacus cucurbitae* Coq. on bitter gourd. *Pakistan Journal of Zoology*, 26: 11-12.

Tan, K.H. and Lee, S.L. 1982. *Bull. Entomol. Res.*, 72: 709-716.

Tandon, R. N. and Verma, A. 1964. Some new storage diseases of fruits and vegetables. *Current Science*, 33: 625-627.

Taneja, S. L., Reddy, K. V. S. and Leuschner, K. 1986. Monitoring of shoot fly population in Sorghum. *Indian Journal of Plant Protection*, 14: 29-36.

Tanigoshi, L. K., Bahdousheh, M., Babcock, J. M., Sawaqed, R. 1990. *Euseius scutalis* (Athias-Henriot) a predator of *Eutetranychus orientalis* (Klein) (Acari: Phytoseiidae, Tetranychidae) in Jordan: toxicity of some acaricides to *E. orientalis*. *Arab Journal of Plant Protection* 8:120-114.

Tayade, D. S. and Chunderwar, R. D. 1977. Occurrence of Aulacophora foveicollis (Lucas), on jowar, Sorghum. Current Research, 5(8):135.

Tenaglia, D. 2006. Missouri Plants: photographs and descriptions of flowering and non flowering plants of Missouri : http://www.missouriplants.com

Thomas, W. 1971. The incidence and economic importance of Watermelon mosaic virus. *N. Z. Journal of Agricultural Research*, 14: 242-247.

Thomas, W. 1980. Watermelon mosaic virus, the cause of a serious disease of cucurbits in the Cook Islands, New Zealand Journal of Experimental Agriculture, 8(3-4): 309-312.

Thompson, A. 1933. Division of mycology. Annual Report for 1932. Dept. of Agric. Strait Settleraents and Fed. Malaya States, Kuala Lumpur (Reports of the Res., Econ., and Agric. Educ. Branches for the year 1932). Bull. 14, Gen. Ser., 53-62. (*Abstract in Rev. appl. Mycol*, 13:216.

Tilavov, T. T. 1981. The daily and seasonal feeding rhythm of the adult of the cucumber beetle, *Epilachna chrysomelina* Fab. *Uzbekskii Biologiceskii Zhurnal*. 8(5): 49-51.

Toivonen, P. M. A. 2003. Effect of Storage Conditions and Postharvest Procedures on Oxidative Stress in Fruits and Vegetables. In: *Postharvest Oxidative Stress in Horticultural Crops*, pp 69- 90. Food Products Press, New York.

Tripathi, G. and Joshi, R. D. 1985. Water melon mosaic virus in pumpkin (*Cucurbita maxima*). *Indian Phytopathology*, 38:244-247.

Tripathi, S. R. and Misra, A. 1991. Population dynamics of *Epilachna dodecastigma* Wied. (Coleoptera: Coccinellidae). *Indian J. Entomol.* 2(2): 203-212.

Trumble, J. T., Moar, W. J., Brewer, M. J. and Carson, W. G. 1991. Impact of UV radiation on activity of linear furanocoumarins and *Bacillus thuringiensis* var. *kurstaki* against *Spodoptera exigua:* Implications for tri-trophic interactions. *J. Chem. Eool.* 17:973-987.

Tsao, P. H. and Ocana, G. 1969. Selective isolation of species of *Phytophthora* from natural soils on an improved antibiotic medium. *Nature,* 233:636-638.

Udalova, V. B. and Prikhod'ko, V. F. 1985. Byulleten VsesoyuzuogoInstituta Gel'mintologii imeni, K. I. Skryabina, No. 41, pp-67-70.

Ullasa, B. A. and Amin, K. S. 1986. Epidemiology of bottle gourd anthracnose, estimation of yield loss and fungicidal control. *Tropical Pest Management,* 32: 277-282.

Upadhyay, G. and Roy, A. N. 1987. Efficacy of certain chemicals in the control of species of *Fusarium* in stored ash gourd. *Pesticides,* 21(5):25-27.

Uppal, B. N. 1934. The adsorption and elution of cucumber mosaic virus. *Indian J. Agri. Sci.,* 4: 656-662.

Utikar, P. G., Mhaske, K. D. and Shinde, P. A. 1986. Occurrence of new marginal leaf blight of bitter gourd incited by *Exserohilum rostratum. Indian Phytopathology,* 39: 492.

Vakalounakis, D. J. and Malathrakis, N. E. 1988. A cucumber disease caused by *Alternaria alternata* and its control. *Journal of Phytopathology,* 121(4): 325–336.

Valcineide, O. A., Tanobe, T., Flores, H. S., Sandro, C., Amico, Graciela, I. B., Muniz, and Satyanarayana, K. G. 2014. Sponge Gourd (*Luffa Cylindrica*) Reinforced Polyester Composites. Preparation and Properties. *Defence Science Journal,* 64(3):273-280.

Van Koot, Y. and Van Dorst, I. 1959. Virus diseases of cucumber in the Netherlands. *Tijdschrift over Plantenziekten,* 65(6): 257-271.

Vande Vrie, M. 1985. Control of tetranychidae in crops. In: *Spider Mites, Their Biology, Natural Enemies and Control* (Eds. Herle W. and M. W. Sebeus). Elsevier Publications, Amsterdam, The Netherlands, pp. 273-283.

Vani, S. 1987. Studies on viral diseases of muskmelon and water melon. Ph.D. Thesis. PG School, IARI, New Delhi.

Vani, S. and Varma, A. 1988. Viral disease problem in muskmelon. Nat. Symp. Pl. Virus Problems in India. IARI, New Delhi (Abst.).

Vani, S., Varma, A., More, T. A. and Srivastava, K. P. 1989. Use of mulches for the management of mosaic disease in muskmelon. *Indian Phytopathology,* 42: 227-235.

Vanisree, K., Rajasekhar, P.; Rama Subba Rao, V. and Srinivasa Rao, V. 2005. Seasonal incidence of pumpkin caterpillar, Diaphania indica (Saunders) on cucumber in Krishna-Godavari Zone. *Journal of Plant Protection and Environment,* 2(1): 127-129.

Vargas, R.I., and Nishida T. 1992 Ecological framework for integrated pest management of fruit flies in papaya orchards. In: Ooi PAC, Lim GS, Teng PS editors. Proceedings of the third International Conference on Plant Protection in the Tropics 20-23 March 1990, pp. 64–69.Malaysian Plant Protection Society, Genting Highlands, Kuala Lumpur, Malaysia.

Varma, A. 1988. The economic impact of filamentous plant viruses: the Indian Subcontinent.

Varma, A. and Giri, B. K. 1998. Virus diseases. In: Cucurbits (Nayar, N. M. and More, T. A. eds). Oxford and IBH Publishing Co. Pvt. Ltd., New Delhi, pp. 225-245.

Varma, B. K. 1961. Bionomics of *Phyllotreta cruciferae* (Goeze) (Chrysomellidae: Coleoptera) reared on radish in India. *Indian Journal of Agricultural Sciences,* 31: 59-63.

Varma, P. M. 1955. Ability of whitefly to carry more than one viruses simultaneously. *Curr. Sci.* 24: 317-318.

Varshney, J. G. 2009. Why weed control. *Crop Care,* 35(10):13-25.

Varveri, C., Vassilakos, N. and Bem, F. 2002. Characterization and detection of cucumber green mottle mosaic virus in Greece. *Phytoparasitica,* 30: 493-501.

Vasudeva, R. S. 1960. The fungi of India (Revised). ICAR, New Delhi.

Vasudeva, R. S. and Lal, T. B. 1943. A mosaic disease of bottle gourd. *Indian J. Agri. Sci.,* 13: 182-191.

Vasudeva, R. S., Raychaudhuri, S. P. and Singh, J. 1949. A new strain of cucumis virus 2. *Indian Phytopathology,* 2: 180.

Vasudeva, R. S., Raychaudhuri, S. P. and Singh, J. 1950. A new strain of *Cucumis virus. Indian Phytopathology,* 2:180-185.

Venkatesha M G. 2006. Seasonal occurrence of *Henosepilachna vigintioctopunctata* (F.) (Coleoptera: Coccinellidae) and its parasitoids on aswagandha in India. *J. Asia-Pacific Entomology,* 265-268.

Verma, A. C. and Anwar, A. 1996. Assessment of yield loss due to *Meloidogyne incognita* in pointed gourd, *Trichosanthes dioica* Roxb. *Afro Asian Journal of Nematology,* 6(1): 92-93.

Verma, A. C. and Anwar, A. 1998. Effect of organic amendments on sprout emergence of pointed gourd, in root-knot nematode, *Meloidogyne incognita* infested field. *Annals of Plant Protection Sciences*, 6(1):102-104.

Verma, A. L. and Anwar, A. 1997. Control of *Meloidogyne incognita* on pointed gourd. *Nematologia Mediterranes*, 25(1):31-32.

Verma, S. K. 2014. Enhancing sustainability in wheat production though irrigation regimes and weed management practices in eastern Uttar Pradesh. *The Ecoscan*, 6:115-119.

Verma, S. K. and Singh, S. B. 2008. Enhancing of wheat production through appropriate agronomic management. *Indian Farming*, 58(5):15-18.

Vijaysegaran, S. 1985. The Occurrence of Oriental Fruit Fly on starfruit in Serdang and the status of its parasitoids. *Journal of Plant Protection in the Tropics*, 1(2):93–98.

Viraktamath, C. A., Tiwari, G.C.; Srinivasan, K. and Gupta, M. 1993. American serpentine leaf miner is a new threat to crops. *Indian Farming*. 10:12.

Virupaksha Prabhu, H., Hiremath, P. C. and Patil, M. S. 1997. Biological control of collar rot of cotton caused by *Sclerotium rolfsii* Sacc. *Karnataka Journal of Agricultural Sciences*, 10: 397-403.

Visalakshi, A., Beevi, S. N., Premkumar, T. and Nair, M. R. G. K. 1980. Biology of *Leptoglossus australis* (Fab.) (Coreidae: Hemiptera) a pest of snake gourd. 5(1): 77-79.

Vishwakarma, R., Prasad, P.H., Ghatak, S.S. and Mondal. S. 2011. Bio-efficacy of plant extracts and entomopathogenic fungi against Epilachna beetle, *Henosepilachna vigintioctopunctata* (Fabricius) infesting bottle gourd. *Journal of Insect Science*, 24(1):65-70.

Vishwanath, S. M., Nariani, T. K. and Chenulu, V. V. 1980. Studies on a melon ring spot virus. *Curr. Sci.*, 49:483.

Vyas, N. L. and Panwar, K. S. 1976. Some new post-harvest diseases of fruits. *Indian Phytopathology*, 29:94-95.

Walker, J. C. 1952. Diseases of vegetable crops. McGraw Hill Book Company, Inc. New York, pp-529.

Walters, S. A., Young, B. G. and Krausz, R. F. 2008. Influence of tillage, cover crop, and preemergence herbicides on weed control and pumpkin yield. *International Journal of Vegetable Science*, 14(2):148-161.

Walters, S. A., Young, B. G. and Nolte, S. A. 2007. Cover crop and pre-emergence herbicide combinations in no-tillage fresh market cucumber production. *Journal of Sustainable Agriculture*, 30(3) 5-19.

Waraitch, K. S., Munshi, G. D. and Chohan, J. S. 1976. A new wilt disease of muskmelon in India. *Curr. Sci.*, 45: 628-629.

Waraitch, K. S., Munshi, G. D., Nandpuri, K. D. and Lal, T. 1977. Screening of muskmelon, wild melon and snap melon for resistance to powdery mildew (*Sphaerotheca fuliginea*). *Phytopath. Medit.*, 16: 37-39.

Ward, C. R. 1985. Blister beetles in alfalfa. Cooperative Extension Service. Circular 536, 9 pp. College of Agriculture and Home Economics, New Mexico State University, USA.

Wasantwisut, E. and Thara, V. 2003. Ivy Gourd (*Coccinia grandis* Voigt, *Coccinia cordifolia, Coccinia indica*) in Human Nutrition and Traditional Applications: In Plants in Human Health and Nutrition Policy, World Revies Nutriton and Dietics Basel, Karger pp-60-66.

Waterhouse, G. M. 1970. *The genus Phytophthora deBary: Diagnoses and figures.* Commonwealth Mycological Institute, Kew, England. *Mycology Papers,* 122: 1-59.

Watson, A. and Napier, T. 2009. Diseases of Cucurbits vegetables. New South Wales Department of Primary Industries. *Primefacts* No. 832. pp-6.

Webb, J. C., Slaughter, D. C. and Litzkow, C. A. 1988. Acoustical systems to detect larvae in infested commodities. *Florida Entomologist*, 71, 492-504.

Webb, R. E. 1963. Local-lesion hosts for some isolates of watermelon mosaic virus. *Plant Dis. Rep.*, 47: 1036-1038.

Webb, R. E. and Bohn, G. W. 1961. A virus latent in some cucurbits. *Plant Dis. Rep.*, 45: 677-679.

Wehner, T. C. and Jenkins, S. F. Jr. 1985. Rate of natural outcrossing in monecious cucumbers. *HortScience* 20:211-213.

Wei, Y., Liu, D. Q. and Zhang, S. 2004 Controlling effects of different insecticides on *Henosepilachna vigintioctopunctata* (Fab.). *China Vegetables*, 4: 39-40.

Welty, C. 2009. Squash vine borer. Agriculture and Natural Resources Fact Sheet HYG-2153-09 [Online]. The Ohio State University Extension. Available at: http://ohioline.osu.edu/hyg-fact/2000/pdf/2153.pdf

Wharton, R. A. 1989. Classical biological control of fruit infesting Tephritidae, *In*: Robinson, A. S. and Hooper, G. (eds.), Fruit flies; their biology natural enemies and control. *World Crop Pests*, Elsevier, Amsterdam, 3(B), 303-313.

White, I.M., Elson-Harris,M.M.1992. Fruit flies of economic significance: their identification and bionomics. Wallingford, U.K: CAB International, 601pp.

Whitson, T. D., Burrill, L. C., Dewey, S. A., Cudney D W, Nelson B E , Lee R D and Parker R. 2000. Weeds of the West. The Western Society of Weed Science in cooperation with the Western United States Land Grant Universities, Cooperative Extension Services. University of Wyoming. Laramie, Wyoming. pp-630.

Wolfenbarger, D. A. and Wolfenbarger, D. O. 1966. Tomato yields and leaf miner infestations and a sequential sampling plan for determining need for control treatments. *Journal of Economic Entomology,* 59:279-283.

Wyenandt, A. and Nitzsche, P. 2006. Diagnosing and managing important cucurbit diseases in the home garden. Bulletin No. E310. Rutgers Cooperative Research & Extension, N. J. Agricultural Experiment Station Rutgers, The State University of New Jersey, New Brunswick, pp-7.

Xiang, C. P., Wu, C. Y. and Wang, L. P. 2000. Analysis and utilization of nutrient composition in bitter gourd (*Momordica charantia*). *Journal of Huazhong Agriculture University* 19:388-390.

Yang, P.J. 1991. Status of fruit fly research in China, p. 161-168. In: S.Vijaysegaran and A.G.Ibrahim (Eds.) *Proc. First International Symposium on Fruit Flies in the Tropics.* Malaysian Agricultural Research and Development Institute, Kuala Lumpur, Malaysia.

Yasuda, K. 1989. Ecology of the leaf footed plant bug, *Leptoglossus australis* Fab. (Heteroptera: Corediae) in the sub-tropical region of Japan. *Proc. of International Symposium on Tropical Agriculture Research: Production of Vegetables in the Tropics and Sub-tropics,* Tsu (Japan), 20-22 Sep 1989, pp. 229-238.

Yoon, J. Y., Choi, G. S., Choi, S. K., Hong, J. S., Choi, J. K., Kim, W., Lee, G. P. Ryu, K. H. 2008. Molecular and biological diversities of *cucumber green mottle mosaic virus* from cucurbitaceous crops in Korea. *Journal of Phytopathology,* 156: 408-412.

Yoon, J. Y., Min, B. E., Chol, S. H. and Ryu, K. H. 2001. Completion of nucleotide sequence and generation of highly infectious transcripts to cucurbits from full length cDNA clone of Kyuri green mottle mosaic virus. *Archives of Virology,* 146: 2085-2096.

Yoon, J. Y., Min, B. E., Chol, S. H. and Ryu, K. H. 2002. Genome structure and production of biologically active *in vitro* transcript of cucurbits infecting zucchini green mottle mosaic virus. *Phytopathology,* 92:156-163.

York, A. 1992. Pests of cucurbit crops: marrow, pumpkin, squash, melon and cucurmber. In McInlar, RGC (ed.) Vegetable Crop Pests. CRC Press, Boca Raton, Florida, pp-139-161.

Young, D. K. 1984. Cantharidin and insects: an historical review. *Great Lake Entomology,* 17(4):187-194.

Zhu, F., Lei, C. L. and Xue, F. 2005. The morphology and temperature-dependent development of *Mylabris phalerata* Pallas (Coleoptera: Meloidae). *Coleopt. Bull.*, 59: 521-527.

Zhu, J. K. 2001. Plant salt tolerance. *Trends Plant Science,* 6:66–71.

Zhu, Y., Dong, Y., Qian, X., Cui, F., Guo, Q., Zhou, X., Wang, Y., Zhang, Y. and Xiong, Z. 2014. Effect of superfine grinding on antidiabetic activity of bitter melon powder. *International Journal of Molecular Science,* 13: 14203-14218.

Zitter, T. A. 1987. Anthracnose of cucurbits. In: Vegetable crops. Fact Sheet. Department of Plant Pathology, Cornell University. pp-732.

Zitter, T. A. 1992. Fruit Rots of Squash and Pumpkins. Cooperative Extension, New York State, Cornell University, Fact Sheet, pp-751.

Web resources

http//:www.ncipm.org.in

http//:www.ipm.ucanr.edu

http//:www.agritech.tnau.ac.in

http//: www.gardeningknowhow.com

http://www.nr.gov.nl.ca/nr/agrifoods/crops/veg_pdfs/cucumbers.pdf

http://www.ca.uky.edu/agc/pubs/id/id91/id91.pdf

About the Authors

Dr. Nripendra Laskar: Presently serving as Associate Professor in Agricultural Entomology and Head, Department of Agricultural Entomology, Uttar Banga Krishi Viswavidyalaya (North Bengal Agricultural University), West Bengal. Dr. Laskar obtained B. Sc. (Ag.) degree in 1996 and M. Sc. (Ag.) degree in Agricultural Entomology in 1998 from Bidhan Chandra Krishi Viswavidyalaya and Ph. D. from Visva-Bharati, Santiniketan, West Bengal. He has 16 years experience in teaching, research and extension experience in the concerned field of specialisation and published more than 30 research papers in different National and International journals of repute. Dr. Laskar authored 2 books, 8 book chapters and guided a number of PG students. In addition he is the founder Joint Secretary of Cooch Behar Association for the Cultivation of Agricultural Sciences (COBACAS), life member of Society for Biocontrol Advancement (SBA), NBAIR, Bangalore, member of the Society of Plant Protection and Environment, OUAT, Bhubaneswar. Dr. Laskar has participated and delivered lectures in 2 International Symposium, 12 National Seminar, Symposium and Conferences so far and presently working on tephritid pests of vegetables and fruits and Indigenous Technical Knowledge (ITK) in pest management.

Dr. Bholanath Mondal: Assistant Professor, Department of Plant Protection, Palli-Siksha Bhavana, Visva-Bharati, completed M. Sc. (Ag.) and Ph. D. in Plant Pathology from Bidhan Chandra Krishi Viswavidyalaya and has qualified NET (ICAR) in Plant Pathology. Dr. Mondal has honoured with 'Young Scientist Award' and 'Best Young Scientist Award' in Plant Pathology. He also received 'Rashtriya Gaurav award' for the contribution of Plant Pathology. Dr. Mondal is the founder Vice-President of 'Cooch Behar Association for Cultivation of Agricultural Sciences' (COBACAS) and founder Assistant Secretary of 'Society of Bio-Resource, Environment and Agricultural Research', editorial board member of a number of reputed journals and life member of different societies. He acted as Chief Editor of the Journal, Green Technology. Dr. Mondal handled/is handling many research projects funded by different Government and Non-Government organizations and published over 30 research articles in the Journals of National and International repute. He is the author of a dozens of book chapters and six books on various aspects of disease-pest management, environment and sustainable agriculture.

Dr. Partha Choudhuri: Associate Professor in Vegetable Crops, Bidhan Chandra Krishi Viswavidyalaya, Mohanpur, Nadia, West Bengal, India. Dr. Choudhuri has completed his B.Sc (Ag) degree with specialization in Hortuicculture and Masters in Spices and Plantation Crops from Bidhan Chandra Krishi Viswavidyalaya, Mohanpur, Nadia, West Bengal and Ph. D. in Vegetable and Spice Crops from UBKV, Pundibari, Coochbehar, West Bengal, India. Dr. Choudhuri has published thirty five research and review papers, six book chapters. He has become reviewer for number of national in international journals. He has handled two research projects. Dr.Choudhuri has acted as external examiner, paper setter for several SAUs.